THE ENCHANTED GARDEN
FLOWERCRAFTS
A MIRROR OF NATURE

THE ENCHANTED GARDEN
FLOWERCRAFTS
A MIRROR OF NATURE

Sedgewood®Press

Published by Sedgewood® Press

For Sedgewood® Press
Director Elizabeth P. Rice
Editorial Project Manager
  Bruce B. Macomber
Production Manager Bill Rose

Produced for Sedgewood® Press by
Marshall Cavendish Books Limited
58 Old Compton Street
London W1V 5PA

For Marshall Cavendish
Editor Pat Sinclair
Designer Caroline Dewing
Production Richard Churchill
Illustrators Sally Holmes
         John Hutchinson

First printing 1988

ISBN 0-02-496880-3
Library of Congress Catalog Card
Number 87-060939
Printed in the United States of America

CONTENTS

INTRODUCTION

Flowers are an important and wonderful part of our lives and our heritage. When we walk around a garden, the medley of colors and the mingling scents of the flowers have an effect that can be both calming and exciting. When we pass a flower stand or florist's shop on a busy or gloomy day, the sight of buckets filled with bunches of colorful blossoms gives the heart – and spirit – a lift.

For natural materials to decorate our homes and set the scene for memorable family and public occasions, we choose flowers – as we do when we want to express our deepest feelings of joy, congratulations or sympathy. Flowers have a special role as a fashion accessory at the most joyous of occasions – such as weddings or festivals – when wearing flowers expresses most strongly what we feel.

Flowers have a special role on our tables too, not only as decorations but as delicious ingredients. In the past flowers were food and drink to people who did not perceive vegetables and flowers as having distinct uses; many people were as likely to toss marigolds in salads and make a dessert from roses as they were to put these flowers in vases.

The sight of a flower garden at its peak is so impressive, it is tempting to wish the season could go on forever and that flowers would never wilt or fade. Happily, this is increasingly possible; more and more plant materials can be dried in the air or in powder – some even in microwave ovens – and will retain their color, shape and form almost indefinitely.

This book is a celebration of the beauty of flowers. It covers every aspect of the craft from the moment you bring flowers into the home to the time when you use them in designs, cosmetics, food and drinks, and when, with a little leisure time to spare, you decide to create everlasting flowers of your own.

Designing with dried blossoms, for example, is a branch of flowercrafts that has lasting satisfaction. A basket of dried rosebuds, marjoram, lady's mantle and larkspur can be pure nostalgia – summer captured as a beautiful and tangible memory. Our instructions show you how easily this dream can come true.

The fragrance of flowers also accounts for a significant part of their charm. Centuries ago people perfumed their rooms with dried flower potpourri and with floral perfumes and cosmetics. Today, when the popularity of natural products is at an all-time high, this is a therapeutic aspect of flowercrafts that is both topical and relevant. Many of our recipes for skin care, hair care and other beauty products draw on medieval herbals for their authenticity; many of our ideas show you how to turn these pure, homemade products into delightful gifts.

Since flowers, from the palest rose to the boldest sunflower, are so beautiful, it would be surprising indeed if they had not inspired the flattery of imitation. Here is a pastime that requires little artistic talent, yet offers an exhilarating challenge. How, with crepe paper, silk or satin, can you hope to emulate nature itself? Our design instructions show you simply and clearly the various techniques needed to create paper, fabric, and even sugar flowers, and our photographs demonstrate clearly just how delightful and decorative they can be.

No matter what your interests or skills, you're certain to find flowercraft one of the most satisfying – and appreciated – forms of artistic expression.

A handful of flowers and leaves picked from the garden combined with a few purchased blooms can make a rich and generous display.

Chapter 1
ARRANGING FRESH FLOWERS

Arranging flowers is very much a matter of creating a feeling, a look, a mood, an impression; and, as an art form, it is a wonderful way of expressing your own personality. No two people would cut an identical bunch of flowers if they had the freedom of a lovely garden, or choose exactly the same blossoms if they were given a free hand in a flower shop.

One person might seek out the most striking shapes of flowers in one particular color and then select others to complement them in terms of tone and texture; another might gather up a modest bouquet of tiny blossoms in pale and insignificant tints; a third person, perhaps without being aware of it, might instinctively select bold, showy flowers in widely contrasting colors; and yet another would concentrate on foliage with well-defined colors and shapes, adding a few eye-catching flowers as highlights.

No two people would arrange a bunch of flowers in precisely the same way. Designing with flowers makes such a personal statement that there are no hard and fast rules to follow. There are, however, certain general guidelines that make it easier to create, with complete assurance, the effect you seek, whether it is casual or classically formal.

Color has a profound effect not only on the flower

Before visiting your local florist, see what plants you have on hand. Potted herbs, a rose bush heavy with flowers, or an evergreen shrub that needs pruning could all provide good display materials.

appear much darker at the base, where it is thicker, but shade to a paler tint at the tip. If the petal is translucent, and light is shining through it, the color will be significantly weakened and appear much paler. This light factor fundamentally alters the character of the color.

ACROSS THE RANGE

Shades, tones and tints are familiar words that have specific meanings in relation to color, meanings that are sometimes obscured in everyday usage. Shades are the deepest hues of any color, created by mixing the raw color with black. To reproduce the color of the very deepest, velvety, old-fashioned roses, an artist would mix red and black; to capture the strongest hue of a purple iris or hyacinth, he or she would need red, blue and black, and so on.

Tones are in-betweens, achieved by mixing the basic color with gray. Red and gray in varying proportions would portray the zingy, vibrant pinks of, for example, carnations, fuchsias and peonies. And at the palest end of the scale are the tints, made by blending the base color with white; red with a dash of white for a red slightly on the pale side of scarlet; white with a dash of red for sweet peas in the most delicate tints of sugared-almond colors.

And there you have it, six colors, a never-ending range of shades, tones and tints of each of them, and a dazzling choice when it comes to the moment of decision; what to put with what in order to make the very best of your plant materials.

Back to the artist's color wheel again. The sequence of the colors in the six segments is the key to the effect each flower and leaf will have on its neighbors in any group you arrange.

COLOR PARTNERSHIPS

To bring the phrase "the attraction of opposites" to life in your designs, try linking flowers and leaves in each of the three pairs of colors opposite each other on the imaginary wheel. Red and green, orange and blue, and yellow and purple are the pairings considered to be most flattering to each other, and are termed "complementary colors."

Suppose you have a single red flower that you would like to use, to become the focus of attention in a design. Help it on its way by surrounding it with green flowers, bracts or leaves. A brilliant deep red carnation nestling in a bowl of that lovely pale green variety of nicotiana (tobacco plant); a cluster of flame-red roses tumbling over a spire of misty Bells of Ireland; even a single rose placed so that it is viewed against a background of leaves on the stem, or that joyful natural gift of winter, lively red berries glowing among the evergreens. No other color but green will pay reds of whatever hue such a perfect compliment. No other neighboring color, and certainly not neutral white, will make the red appear so vibrant, yet allow the eye to take in every detail of the line, form and texture of the flower or fruit.

At first glance, the orange and blue partnership may seem to offer less spontaneous possibilities. But the theory holds good. Gather together a handful of orange and blue pansies, orange marigolds and blue cornflowers, or a single sunshine gold alstroemeria amidst a bunch of midnight-blue lilies and you will find that they sit in perfect harmony.

For the third "color marriage," yellow and purple, there can be no more delightful examples than a

Above: *This predominately yellow and white arrangement, using garden-grown wildflowers, comes to life with the addition of flowers in the complementary color of purple.*
Opposite: *Simple but novel, this design, with anemone and spray chrysanthemum heads on a wire rack, uses a similar color combination.*

child's Mother's Day bouquet of dainty primroses and violets, or the showier springtime alternative of daffodils and anemones.

If a single flower has survived a bunch that was an extravagant splurge at the florist's, turn on the spotlight by mingling it with yellow flowers. The purple will glow like a flower bathed in sunlight.

Just as an experiment, and to prove the point, hold a flower of a similar color against blues and reds. These are the colors placed on each side of purple on the color wheel and, being so closely related, want a share of its limelight. Now you will scarcely notice your purple flower at all. It has become a "shrinking violet."

GOOD NEIGHBORS

Color mixtures chosen from hues that are neighbors on the color wheel are termed "adjacent." A bunch of anemones is a perfect example; all the blue, purple and red flowers attract equal attention and form a unified group. Another would be orange and yellow spray chrysanthemums arranged with green foliage of a more or less equal hue (dark green with strong shades, light green with pale tints). Don't expect there to be any obvious stars of the show, just a

Far left: *A single red chrysanthemum in the midst of muted green foliage provides a strong focal point for this arrangement.*
Left: *By contrast, red flowers scattered among salmon and pink colored blossoms lose their dramatic impact, but add depth to this vase arrangement.*

uniformly good-looking chorus.

In many flower designs green, the color of such a high proportion of foliage, joins gray, brown and white as a neutral color. This is because, in the context of plant materials, we expect to see green leaves (brown ones, too) and so tend to discount their color as a contributory element.

MATCHING MOODS

So much for the effect that various colors have on each other. What about the effect they have on you, and on your home? You can match your mood, set the background to a party or other special occasion, emphasize good architectural points such as an arch or window or even disguise less agreeable features simply with the colors of the flowers you choose.

Red flowers have a number of connotations. First and foremost, perhaps, is the romantic one. Red roses signify true love; red rosebuds a young, tender, unfolding love. A heart-shaped bouquet of red roses offered on Valentine's Day or on a special anniversary speaks volumes, and does so more charmingly than words. (Other examples of the enchanting – and largely Victorian-inspired – language of flowers are given on pages 98-99).

There is no problem if you wish to create a romantic mood, then. Red flowers of other species can convey a bold, devil-may-care spirit. A pitcher of bright scarlet chrysanthemums might seem to suggest an uncompromising attitude, almost a challenge, and a jug of field poppies a relaxed, informal mood. Casually blended with a cluster of grasses or cereals, poppies look delightfully at home in a farmhouse-style kitchen, setting the scene for a fall party or a simple meal of soup and salad.

If reds span every mood from high romance to happy-go-lucky, pink flowers are without exception

in a softer mold. Pink roses, carnations and pinks are the number-one choice of young brides for flowers to carry and wear, and set a nostalgic scene in formal arrangements in church or at the reception. A basket of pink daisies to greet the birth of a baby, a cascade of pink lilies as a gift for an honored guest – somehow pink spells achievement and glamour.

Orange flowers positively glow. Arrange a basket of orange gaillardias, or gerberas in the hearth when it is not quite cold enough to light a fire, and you can almost toast your hands on the comforting effect. Fill an earthenware pot with pale tangerine-orange wallflowers or marigolds and they will gleam beside a patio barbecue long after the last embers have died away. To bring out every last flicker of warmth, arrange orange flowers, from elegant lilies to innocent daisy types, in copper pots and pans.

SUN POWER

There is no doubt about it, yellow is a happy, "open the windows and let the sun stream in" color. You have only to introduce yellow flowers of any kind and you create rays of imaginary sunshine in the home. It is, I suspect, impossible to arrange yellow flowers and yet create a somber design. From the first citrus yellow daffodils of spring to the rich, golden tones of falling leaves and, timed to coincide with them, glowing dahlias and chrysanthemums, yellow flowers gladden the heart. In between there are dazzling buttercups and yellow-centered daisies, so tempting that creative fingers itch to twine them around a hatband, or link them to form a bracelet. Canoeing on a placid mountain lake, an exuberant Fourth of July picnic – that's the kind of mood that the yellow flowers of summer seem to suggest.

Think of green flowers, of green foliage, and you can almost hear the tinkling of a waterfall, the swish of a stream, the spray of a fountain. Green has a cooling effect. As a neutral color in the plant context, it separates vibrant colors and dilutes exuberant

Partly smothered by a veil of Queen Anne's lace, the showy, vibrant pink gerberas in this display become more pleasing to the eye.

mixtures. In this way the addition of green foliage or other plant materials renders sharp color combinations attractive and acceptable.

Blue is for beauty, it is said. It is also a "cool" color but – unlike green – not considered a neutral hue that merges into the background. From the palest tints of dainty flowers such as speedwell to the deep, resonant shades of grape hyacinths and iris, blue spans almost every mood.

Reminiscent of a brilliant cloudless sky, blue flowers evoke a countrified air; perpetual cornflowers, forget-me-nots, periwinkle, even thistles look bright and beautiful against a glowing orange or yellow wall or curtain. Tall spires of midnight-blue delphinium, larkspur and clarkia, and their less stately cousins, lilac, flag irises and lilies, lend an air of unmistakable elegance. These are a grand choice for formal occasions and large areas.

Purple, perhaps the most impressive and formal color of all, is not all that easy to use. Clusters of purple flowers in a group composed largely of yellow, blue and white can be the most rewarding. But having said that, a shallow bowl filled to overflowing with a single variety of deep, shadowy purple iris can look positively regal.

WHITE LIGHTS

The mention of white as neutral in terms of plant color may be misleading, and is worthy of comment. Some avid floral artists can scarcely conceive of a design without highlights of white – and that is just the element it does provide. Do not blend white flowers with others that you want to emphasize – they will surely steal the show – and never dot white flowers here, there and everywhere in a design, unless you crave a polka-dot creation. For almost without exception the white material will glow out of the design like a beacon.

It takes exceptions to prove a rule, or disprove a statement. In the case of white plant materials, the exceptions are all the misty, frothy flowers. Tiny gypsophila (baby's breath) and the alternatives of snowflaky wild carrot (Queen Anne's lace) or wild chervil can draw a soft, flattering veil over powerful colors and strong outlines, and create the illusion of a soft tint in place of a vibrant hue. If ever someone gives you a bunch of carnations or pinks that are too bright for your liking, curtain them with see-through whites – the floral equivalent to curtain "sheers" – and give the showy creatures a new and more modest personality.

Juggling successfully with flower colors, shaking the floral kaleidoscope, is almost like having a new source of light at your fingertips. Use white, yellow and the palest tints of pink, orange, green and blue flowers and foliage to advantage.

These are the colors to brighten a dark corner, draw attention to a shadowy arch or other design feature, silhouette against a dark piece of furniture – a polished dark oak chest, maybe – or use outdoors for an evening party. As dusk falls and trees and shrubs fade into gentle obscurity, a tub of pastel-colored flowers will glow far into the night.

EXTRA REFLECTIONS

By contrast, strong, bold colors will catch the eye only if the design is adequately lighted. In shadowy situations even the most brazen reds and oranges can seem gloomy, and deep blues and purples will completely disappear.

It follows, naturally, that light has a visual make-or-break part to play in floral crafts. Most flattering of all, and difficult to imitate precisely, are shafts of streaming sunlight. Yet even this can prove to be too much of a good thing. For the sake of your display's longevity, place fresh and dried flowers – whichever you prefer – in diffused light, or where light is reflected, but not in the full glare of hot sun. Delightful as it is to see a jug of flowers on a sunny windowsill, the pleasure can be short lived. Sun and strong light draw out color from natural plant materials and, of course, from synthetic materials too.

You might try turning a spotlight on a flower group. Experiment by twisting the light source so that it strikes the flowers from different angles; you will find that you can completely change the mood of the design. Shine a light from behind, through a glass bowl filled perhaps with glass marbles and holding delicate foliage, and you create the illusion of cool, running water. Throw a light onto a well-defined flower arrangement against a pale wall, and the shadows it casts will add a new dimension. Light a bowl of flowers from above and you look into a sea of upturned faces, like pansies basking in the sunshine. Light a group from one side and you create an apparently wider range of tints and shades in the plant materials.

Color and light. Light and dark. A single flower color – a monotone design that is anything but monotonous – or a group composed of all the colors of the rainbow. Which is it to be? It's an exciting, exhilarating, never-ending choice.

GETTING INTO SHAPE

The long and the short of it is that you can arrange flowers in any shape or form that pleases you. For what are flowercrafts except an art form, a way of expressing personal taste and individual interpretation of color and balance? When all is said and done the only judgment to be made is whether the casual cluster of flowers in a jug or the elaborate arrangement on a concert stage is pleasing. For beauty, as everyone knows, is in the eye of the beholder.

Inspiration for flower designs is all around us, in gardens and many other kinds of landscapes, in the work of artists, designers and photographers, in glimpses of color, tricks of the light, in anything, in fact, that captures our imagination.

NATURAL BEAUTY

Nature itself is, of course, the greatest inspiration of all. If we can arrange flowers so that they seem as "natural" as possible, and appear as close as can be to the way they grow, then our designs cannot fail to be attractive and pleasing.

As we have seen, a close observation of the countryside and the urban garden gives a valuable insight into the infinite range of colors that are all around us. Such observation also opens our eyes to the shapes formed by plants as they grow. We notice the verticality of some strong stems like gladioli and bulrushes; the severely angular twists of others, a gnarled old apple tree, for example; the gentle curves of delicate grasses gradually bent in the path of the wind; the way the branches of a sturdy oak tree radiate from the tree trunk like water cascading from a single source – and all these factors contribute to the ways we choose to arrange flowers.

Like slender, long-necked birds scanning the horizon, these freesias attract attention with their simple forms and clear-cut lines.

EXPLORING THE ARTS

We can draw our inspiration from the work of artists, observing with what simplicity Van Gogh treated a pot of sunflowers, and the glorious profusion of the flower collections of the works of the Dutch Old Masters. The latter painted spring, summer, autumn and winter flowers as they bloomed and gathered them together later, in multiseasonal bouquets on canvas. Today we can use dried or artificial flowers in the same way.

Visiting art galleries to see how painters over the centuries have portrayed flowers, museums to study floral motifs on pottery, jewelry, drapes and tapestries, and exhibitions of photography for a through-the-lens interpretation of plant materials, allows anyone who loves flowers to gather a series of artistic impressions to interpret, however freely, for their own designs.

Armchair explorers, with no time for gallery browsing, can build up an inspirational collection of postcards, greetings cards, magazine cuttings and scraps of material. A Greek holiday postcard showing an earthenware pot brimming over with sweet basil set on a flight of sun-drenched white stone steps does not necessarily make you want to rush out and buy a packet of basil seeds, though it well might. The soft, cushiony shape of the plant sets the mind racing. Could you achieve a similar effect with a bowl of pale green nicotiana flowers veiled with misty alchemilla mollis and displayed against a dazzling white wall? The answer is probably. But you don't quite know. Until you try.

The natural twist and flow of stems and the rounded and angular shapes of flowers and leaves are infallible clues to the mood and feeling of the arrangement you can create; evoking the bouncy, soft characteristics of basil would be impossible with hard-edged mahonia or holly leaves, or with angular flowers like lilies and birds of paradise.

THE TRADITIONAL SHAPES

Whether you plan your designs along traditional lines – to produce arrangements that could lead eventually to show and competition work – or decline to follow any guidelines at all is purely a matter of choice. But it is always worth learning a little of the recognized floral art code in order to create arrangements with "formal" outlines.

Traditional floral art has its roots firmly planted in the close relationship between the shape of the plant

A plastic saucer holding a cylinder of Styrofoam supports this pyramid of daffodils.

material and the shape of the arrangement – a tip well worth taking, however much freedom you like to give yourself. Traditional design shapes may remind you of a geometry lesson – there are equilateral triangles, asymmetrical triangles, L-shapes, curves, verticals, horizontals, and crescents – but interpreted in flowers and flowing leaves, these shapes are far removed indeed from mathematical models.

The classroom and calculator do come to mind, however, when figuring proportion and balance. In formalized designs the length of the longest stem is carefully calculated to be in a predetermined ratio to the height or width of the container, or in some cases the base on which the container stands.

In a horizontal design, the long, low kind that looks so much at home on a dining table, the main side stems are generally thought to be in perfect visual balance when they are each one and a half times the length of the container. Similarly for a vertical arrangement – the type that so eye-catchingly fills a niche or alcove – it is recommended that the tallest stem should extend to a height one and a half times that of the container. It is not a question of taking a ruler to every arrangement, but of judging the

proportions by eye. Time and again you will find that what looks most easy and natural in this type of design turns out to be within these guidelines.

It must be emphasized that we are dealing only with traditional designs that verge on the formal. Outside these limits, it is perfectly possible to place a bunch of white marguerites in a tall cylindrical vase, with no stems showing at all, and achieve a design that will have an agreeable vertical shape.

One of the simplest "geometric" shapes is the equilateral triangle, which can be created on a flat dish, atop a tall container or, most formal of all, and often referred to as the "English" design, on a pedestal. Draw an imaginary line through the center of the container and the planned design, and that is where you place the tallest stem. The height of this stem should relate to the size of the container – it would look awkward to have a very long upright stem "growing" out of a small compact container.

CARDINAL POINTS

The central or main stem is the key to all traditional floral art, and should be chosen most carefully. If you are aiming for crisp, straight lines, select a strong stem that will not wilt under the strain after a day or two. For a softer look, choose a stem with a very gentle curve.

Depending upon what effect you wish to create, the three points of the triangle may be defined either with bare twigs for a stark silhouette; or with flowering shrub branches such as forsythia and winter jasmine; or with spires of flowers such as delphinium, larkspur, clarkia or Bells of Ireland. For large pedestals, slender branches of foliage of lime, beech, copper beech or oak look good, and will evoke the forest gracefully in your work.

If your triangular arrangement is raised in any way – if you are creating a table arrangement on a pedestal cake stand, or using a floor-standing pedestal – it is important to select lower side stems that have natural curves, some bearing to the right and some to the left. Be careful about snipping indiscriminately with the pruning shears, since the ultimate choice will be between trying to twist some stems almost to the breaking point, or placing half of them the reverse way around, revealing the backs of the leaves. No matter how attractive that view is, of one point you can be certain – the two sides will not match.

Once the central stem is positioned, the base of that stem becomes the "growing source" of much of the material to follow. So be sure that there is plenty

Above: *Under this arc of flowers lies a foam block kept in position by a jar lid.*
Right: *For an alternative display in restricted space, try an L-shaped composition keeping the plants free flowing, and not too severely geometrical.*
Opposite: *A plastic ring filled with foam supports this brilliant, multicolored flower ring.*

of stem-pushing space around it. When you have just a few more vital stems to place, it is frustrating to discover that the holder is crammed to capacity or the block of foam is overcrowded to the splitting point. (We go into more detail about these technicalities and how to avoid such frustrations in the following chapter.)

In a well-balanced design, the eye should travel from the tip of the central stem down to the base of the stem, and then out to each side. It is usually best to "strengthen" the central stem visually by placing another one slightly shorter and just to one side of it, partially overlapping, and another stem, shorter still, slightly to the other side. This avoids a spindly look near the center.

FOCUSING ATTENTION

Since all other stems are placed as if they spring from the same source – like the branches of a tree radiating from the central trunk – this area becomes the center of attention. In floral art terms, that is to say that it

becomes the focal point. This is where you place your choicest flower, a single blossom of contrasting color, a cluster of light-catching berries, or a bouquet of tiny blooms nestling among long-stemmed flowers – whatever you choose to make a visual statement.

From the central "growing point," place the two main side stems, which in an equilateral triangle must be of equal length, and then strengthen them with other stems that are slightly shorter. Now you can begin to fill in the design, recessing some flowers close against the container, and thrusting others forward to avoid a flat, one-dimensional arrangement.

As you fill in the outlines, keep the visual balance equal. This does not mean placing identical flowers in an identical spot on each side, or creating mirror images. It simply means having an equal "weight" of material. To achieve this end, it can be helpful to divide the main flowers and stems into two groups; then you will be sure not to end up with a lopsided look.

For an asymmetrical triangle (which might, in strictly nonmathematical terms, be described as a cross between an equilateral triangle and an L-shape), follow the same guidelines but place the main upright stem off-center and make one side point markedly shorter and a little higher than the other.

An L-shaped arrangement is a perfect example of just how well balanced an asymmetrical design can be. The vertical stem should be about one and a half times the length of the container. The horizontal stem, placed almost parallel to the table or window ledge, will have more visual emphasis than the upright one, and appear longer than it really is. Compensate for this optical illusion by cutting the horizontal stem to between a half and three-quarters of the length of the upright stem. For instance, if the container is a dish, 8 inches long, then the vertical stem should measure 12 inches and the horizontal one 6-9 inches.

Do not feel compelled to take the capital "L" shape too literally in the interpretation of this design. A harsh right angle looks most ungainly. Place two or more shorter stems behind the main upright and leaning slightly backward and, still behind the vertical, at least two of the largest flowers, leaves or pieces of fruit to balance others enclosed within the angle. Without these additions the design not only looks too angular, but also seems imbalanced and in danger of toppling over.

BEAUTY RULES

Traditional flower arrangements are not all sharp angles and ruler measurements, of course – there is plenty of scope for gentle curves, smooth arcs, natural swoops and swirls forming crescents and even S-shapes. One of the most classic shapes of all is the Hogarth curve (also known as the "lazy S"), named after the 18th-century English painter William Hogarth, who called all flowing curves "the line of beauty." The elongated S, created with naturally curving stems, sits gracefully on a raised container such as a candlestick, the plant material curving gently upward on one side and downward on the other – though it is not an easy balance for beginners to achieve.

A similar container, one with a built-in pedestal, can be used to create a formal crescent shape, an arc of stems forming a dome effect in the center and splashing down on each side like spouts from a fountain. An informal no-rules-to-follow version could simply be a bunch of tulips dipping in all directions over the rim of a tall vase.

An alternative is to invert the design. This scheme,

with the sides curving upward in a continuous gentle arc, can be balanced on a raised container. It will also look coolly elegant in a shallow glass bowl with icy blue and green material. You could use spiky stems of broom to outline the shape, and fill in with green hellebores, blue anemones and forget-me-nots.

Above: *Lily-flowered tulips held securely in a cylinder of foam within a shallow container create a sculptural, diagonal line.*

Opposite: *More loosely defined, the shape of this bunch of pinks, veering off-center from the vase, provides an unusual feature.*

NATURAL DESIGNS

In all types of floral art, the size and shape of the flowers, leaves and other plant materials used is just as important as what might be called the mathematical proportions of the design. The main central stem of an equilateral triangle may well be in correct proportion to the size of the container, but it would look uncomfortably top heavy if it had a giant dahlia perched on top of it.

In general, to achieve an easy-on-the-eye balance in traditional arrangements, flowers and foliage should be placed in graded formation. For the points of a triangle, the tips of an S-shape or the limits of a crescent, use the thinnest, finest stems, or flowers in tight bud. In the center of the design, place sprays of medium-sized leaves and moderate-sized flowers or those in half-open bud. For the base of the design you can achieve visual weight with large leaves such as those of hostas, bergenias, geraniums, nasturtiums – depending on the scale of the group – and the largest of the flowers, with fully opened blooms.

This sequence is entirely logical, as it follows the natural growing pattern. Consider a stately stem of delphinium just coming into flower. The buds at the top are still tightly closed, forming a point like that of a sword. Those toward the center of the stem are beginning to open, and the shape is fuller and more rounded. And those at the base of the stem reveal the promise of things to come, the open flowers in all their glory.

GARDEN EXERCISE

It is an interesting exercise, relaxing, too, to walk around a garden or nursery, mentally putting the flowers and foliage into pigeonholes, into the three categories of pointed, midway and rounded – for the tips, center and base of stylized designs.

A clump of purple foxgloves just coming into flower invites transition into the three points of a triangle. Hollyhocks, too large to use on the stem, make pretty fillers at the base, each flower used separately. Trails of variegated periwinkle and lamium leaves make pretty green and cream cascades, while honeysuckle and clematis offer interesting foliage. Euphorbias bridge the gap between flowers and foliage with their brilliant green heads.

In the following section we go into the intriguing hunt for unusual containers, for formal and fun designs, and explore the many ways of keeping stems firmly in line.

CHOOSING CONTAINERS

No dictionary definition could possibly do justice to the description of a container for flowers. That fascinating article can be anything you can imagine or use to put flowers in, on, on top of, behind, beside, through, around or under. And, even if it is for fresh flowers, it does not need to hold water. That's where an inner container comes in. So when it comes to choosing something to complement and contain fresh, dried or artificial flowers, one thing is certain. Absolutely anything goes.

Just to get the sense of the numerous choices around us, let's have a look at the possibilities. You can arrange flowers in a ceramic pitcher inside a braided bread ring as the centerpiece for a Thanksgiving dinner; on a woven cane tray; balanced elegantly, pedestal fashion, on top of a tall wine carafe; behind a twisty, twirling piece of driftwood; tucked beside a scalloped piece of blue coral; pushed through the criss-crossed wires of a cooling rack; perhaps ringed around the rim of a silver candlestick; or even recalling a Victorian bell jar, under an upturned glass tumbler.

There is obviously a close link between the container and the shape of the arrangement you plan. Traditional designs usually demand containers that facilitate building the triangles, horizontals and crescents we discussed in the previous chapter. But fun and even fantasy enter into the proceedings when you are creating free-style designs. Often a distinctive container will call for the simplest floral treatment. A single white lily in a piece of steel piping might be enough to say it all.

Cookie sheets, teapots, molds, storage jars and kitchen canisters are just a few of the potential containers you may find about your home.

A Delftware container complements the blues of the anemones and grape hyacinths.

stand obediently in line. If you fill the neck of the vase with a piece of crumpled wire mesh, creating a jagged dome full of uneven holes, you can push stems in at a variety of angles – confident that you know who is the boss.

By filling the aperture with a piece of stem-holding foam – presoaked in water for fresh materials – extending a little above the rim, you can angle stems horizontally and almost vertically downward. They will all be held firmly and (if it applies) be able to take up water from the foam.

A neater way by far, however, is to attach another container to the top of the vase, and fill this with foam. Keep it topped up with water as required. An ideal solution for a cylindrical vase might be a plastic "foam saucer" that has an indentation just the right size for holding a cylindrical block of foam. Stick the saucer firmly to the vase with extra-tacky florist's clay or water-resistant tape, wiggle it relentlessly to make sure it is secure and then go ahead with your arrangement, on a simple, improvised pedestal.

For flower displays on a large scale, such as a true, floor-standing pedestal, stem-holding precautions have to be more rigorous. As the container will never show, it does not matter how mundane and utilitarian it is. A large, deep vessel such as a pâté or baking dish will do. Fit it with one or two rectangular blocks of soaked foam standing on end and extending well above the rim. This will accommodate downward-sloping stems. Then, to make sure that a concentration of thick stems does not destroy it, cover the foam with crushed wire netting. Secure the dish to the pedestal top by the surest means you know – wrapping it around and around, like a parcel, with fine twine, wiring it to the pedestal rim or the upright.

THE CONTAINER HUNT

One of the most exciting aspects of flowercrafts is the search for "the right container." Indeed, the continuing hunt for classic or unusual items can be an absorbing pastime in itself. Once you are really smitten by the bug, it is almost impossible to let a garage sale, a flea market, thrift shop or auction go unheeded. You never know what you might find.

A lovely rose-spattered coffee pot that is chipped or cracked may be unable to take its place on the dinner table. But that does not stop it from looking absolutely stunning, spilling over with clusters of old-fashioned roses, any one of which will discreetly cover the flaw. A two-handled blue and white china soup tureen may have outlived its lid but be ready for

The container and the plant materials naturally forge a strong visual link. And in many flower designs there is another link that is anything but visual – the holding material that keeps the stems firmly in place and, according to its type, allows you to create any shape you wish.

STAGING THE PROPS

Take a tall, wide-necked cylindrical vase. Using no hidden aids (known in the floral art world as "mechanics") at all, you might arrange a handful of flowers gracefully, but find you have little control over the stems. Some may decide to flop obstinately this way and that. By securing a holder in the base of the container, you can be sure that even the heaviest stems – such as branches of apple blossoms – will

a new lease on life as an elegant container for a symmetrical arrangement. A Victorian oil-lamp base may cease to cast a romantic glow on life, but that base, as shapely a pedestal as you will ever find, can gleam anew with a profusion of softly-trailing flowers and leaves. A wineglass with a twist stem may be somewhat depreciated in value if it is the sole survivor of a set, but it can look enchanting with a bouquet of wildflowers on a dressing table.

If your inclination in floral art is toward classic and historical designs, contemporary, period containers can be costly. Blue and white Wedgwood urns, alabaster cupid vases, Venetian glass epergnes and hand-painted porcelain bowls are very often expensive. But florist's shops and flower-arranging clubs have a wide selection of passable reproductions at affordable prices, and it is well worthwhile starting to build the foundation of your "flower room" shelf from here, and then go bargain hunting.

JUG ARRANGEMENTS

Jugs of all kinds are natural containers for flowers, and are ideal for many different kinds of arrangements. If you can find one of those beautiful two-handled jugs – they are available in various colors – you can create a whole medley of symmetrical designs from equal triangles to groupings that are perfectly balanced. Forget all about symmetry if you are using any vessel with only one handle, such as jugs, drinking mugs, cups, coffee pots and teapots; the handle will throw any carefully contrived design out of balance.

For a lovely rough and tumble effect, you can leave all the holding materials on the shelf and just let the stems have their own way. A yellow jug brimming over with white ranunculus, their yellow centers perfectly echoing the golden container; a blue and white jug amassed with white and blue anemones or a deep green jug filled to overflowing with dainty lilies-of-the-valley and their leaves – such combinations can create some delightful effects.

Think of the container and the flowers as two partners in design, and try to relate the colors in some way. A plain whiter-than-white container holding only brilliant red, blue or orange flowers can be quite distracting, the blooms appearing to float on nothingness. Add a cluster of white flowers, at least one tipping the rim, and the whole group comes together.

Study patterned containers carefully before you make your flower selection, and be aware of the need

A china teapot with just a hint of decoration is the perfect container for this showy bunch of sweet peas.

to establish a visual link. A blue Oriental porcelain jug spattered with cherry blossoms and chirping birds, for instance, does not demand a slavish copy; you don't have to wait for blossom time to come around. Most blue flowers will look in their element, though a very deep shade – some hyacinths, for example – might be top heavy. A mixture of pink and blue anemones would spring no surprises, but could look messy atop a busy decoration. Pink with a single splash of blue, or blue with a dash of pink, could be more restful.

There is something about the rustly papery quality of helichrysum that marries particularly well with the Oriental image. Thick, full white *Helichrysum bracteatum* curving this way and that way on their gawky stems could steal the show. And for fun and frivolity add a lone highlight, such as one pink freesia.

GROUPING FOR DESIGN

Anyone who loves to have flowers in the home is almost bound to build up a collection of jugs. You find a lovely old stoneware jar one time, a tiny creamer the next, add another, and another, and suddenly you are faced with a choice. Which one to use as a container today? It is a choice that need never be made. You can create a fascinating still life by grouping three or four jugs together and linking them visually with flower types and colors. It is a mix and match situation with endless permutations. Stagger a group of jugs in order of size along a dining table, and they are great for a casual party. Arrange them against a wall that sets off the colors to perfection. Cluster them in the center of a side table. Or create a miniature vision of small jugs filled with tiny wildflowers – lawn daisies, forget-me-nots, speedwell, primroses – to attract a bird's-eye view on a low coffee table.

From the minute to the magnificent, you can have floral fun with one of those massive water jug and hand basin sets that sometimes appear at flea markets. You can play it straight, fill the neck of the jug with crumpled wire mesh and go to town on a striking display. A jug with pink overtones would look pretty with a profusion of pink and green hydrangea flowers – at their best when they are just beginning to open – peonies, roses and (a personal favorite) alchemilla mollis.

Stand the jug of flowers in its matching bowl and you can add an amusing extra. Tape a piece of soaked foam just inside the bowl, close to the rim, and fashion a complementary design of smaller blossoms – side shoots and buds – and trailing leaves, some dipping low onto the table, some climbing like tendrils to link the jug and bowl.

Arrangements for the lunch or dinner table look

Dog roses team well with this metallic creamer.

comfortably at home in one or two pieces of the matching tableware. A cup and saucer with a trailing design in the cup and a few flower heads floating in the saucer; a similar treatment for a gravy boat or stand; individual floral tributes at each place setting, in egg cups or custard cups from the set, guarantee a look of complete harmony throughout the meal.

GLASS AND MARBLES

Glinting, glittering, glowing glasses make fabulous containers for mealtimes, at their best when paired with candlelight. A single glass holding a few flowers might not make much impression in the center of a large table. But you could set one beside each place. Have a linking theme but a slightly different color emphasis in each arrangement, and the notion can become a conversation ice-breaker at the start of a meal.

Trumpet-shaped wineglasses can pose a problem for the arranger, the wide V neck offering the stems too much freedom. A small piece of crumpled wire would hold the stems and be lost, visually, among them. But this is a solution only for junk-shop items. Never risk scratching valuable glass in this way. A handful of tiny smooth pebbles or marbles is one answer. Another is to insert a criss-cross mat of short, clean stems to get to grips with those in the arrangement.

There is a certain coolness and tranquillity about glass containers that suggest special effects, particularly those relating to water. Partly fill a goldfish bowl or a rectangular glass bowl with marbles and go for a casual look with a bunch or two of head-drooping tulips. The waxy sheen of their petals perfectly picks up the cool cue.

Polish up a heavy glass dish – even an ashtray would do – and be sure it is absolutely dry before you attach a holder slightly to one side. Arrange a few stems in graded heights – let's choose lemon alstroemeria for their ice-maiden looks – cover the metal holder with concealing leaves and trail a few flowers and leaves, letting some dip into the water.

MAKING FLOWER FLOATS

Floating flowers, a tradition borrowed from the East, are another choice. Select a shallow glass or china bowl, fill it with water and cut off the flower heads with a minimum of stem, just enough to leave them with a drinking straw. At its simplest you can float a single perfect flower – a cream spider chrysanthemum, a pale pink peony or a scarlet camellia, maybe.

It is fun to be more adventurous and create a flower patchwork. Make rings – concentric circles – of yellow doronicum and white marguerites with a single yellow rose in the center. Pick off the lower blooms from a stem of blue delphinium and ring them around pink and white campanula flowers – the possibilities are endless.

On a similar theme, place a wire cake-cooling rack over a shallow glass baking pan of water and arrange short-cut flower heads in a geometric or abstract pattern, just pushing them through the holes. It is like painting a canvas with living color, and use of this device ensures that the flowers do not float out of alignment.

In the kitchen you may unearth a surprising array of unusual containers. Muffin tins could take single flowers or tiny nosegays in each of the six holes; you could use a metal coffee filter cone with an inner container, a fluted brioche mold or ice cream bombe, as decorative as many a custom-made vase; a brightly colored enamel colander (lined with plastic to hold a block of foam); a cane cutlery basket for a Thanksgiving theme of cereals and fruits; a cast-iron skillet (protected against the moisture) spilling over with gazania, marigolds, geum and nasturtiums in sunny-side-up breakfast egg colors – you can turn your kitchen equipment into a versatile battery of flower containers.

FRUIT OF THE WOOD

Wood is a splendid background for flowers. On your treasure hunt in the kitchen, do not discount bread boards, chopping boards, cheese boards and rattan trays just because they are flat. A spot of tacky clay will hold a dish or saucer in place. In goes a holder or cylinder of foam, and you are off and running. The tempting combination of flowers and fruit is particularly effective when you use a wooden container, whether it is a board, an olive pestle and mortar, an old dairy butter mold or a wooden canister.

Fruit offers you a whole new range of shape and texture contrasts – no flower is quite as round, quite as shiny as a well-polished apple. Fruit also poses different attaching problems, because of its weight, and in most cases needs to be wired in place. You will need a package of florist's stem wires or a reel of medium-gauge wire. Twist a wire firmly around each short stem of small russet-colored apples and pears, or push a wire through the fruit at the stalk end and twist the two ends together to make a false stalk. That is how to cope with tangerines and the other small "oranges," limes and lemons. Wire clusters of cherries together into a voluptuous bunch. And weave a long piece of doubled wire around the stalk through a bunch of grapes, twisting it into a firm-gripping loop at the end.

You may find it helps to prepare a container with soaked foam and then cover it with a high-peaked dome of crumpled wire mesh. Wire or tie the netting to the container so that even the heaviest fruits cannot dislodge it, and then twist the fruit wires around the dome. Tuck in soft and pretty flowers and leaves in a design to admire now and eat later.

Perhaps it is stating the obvious, that vases come in all shapes and sizes, to suit all moods and occasions, in all materials and textures. But there is a wealth to choose from – for instance, a '30s-style pink shell-shaped wall vase to fill to overflowing with sugared almond-colored ranunculus, a rough- glazed pottery cube section – a splendid foil for bleached seedpods; a porcelain specimen vase to hold a single rosebud or camellia; or a green glass urn in which to create a masterpiece of foliage, fruit and flowers.

Under a lavish display of fruit, flowers and foliage, soaked foam covered with a cage of wire mesh ensures that all wired stems will stay firmly in place.

35

USING AROMATIC PLANTS

The heady fragrance of night-scented stock and nicotiana at evening time, the beckoning waft of deep perfume from damask and moss rose bushes, the sweet, almost moist aroma given off by a carpet of thyme as it is crushed, and the culinary-inspiring spiciness of fennel, marjoram, mints and dill weed as you brush past them in the herb patch are delights that linger long in the memory – delights that you can transport into the home, combining the pleasure of the color and texture of flowers and foliage with their heavenly fragrances.

There is nothing new, of course, in the notion of perfuming rooms with aromatic plants. In the Middle Ages fragrant plants were strewn on floors and even in the streets to mask offensive odors. Herbs were hung in bunches to discourage insects from making their homes in kitchens and living rooms – in wardrobes and linen chests, too – and in sickrooms as gentle antiseptics. Churches and public halls that may not be in constant daily use can be transformed by a fragrant arrangement. It is a strange thing, but all human senses are so closely linked that a pleasing aroma in a room can make it seem lighter, brighter and an altogether more agreeable place to be.

PLANNING AHEAD

If you are arranging flowers for a big occasion, and if the timetable allows, complete the designs the evening before the event. Make sure that the containers are well topped up with water, spray all the flowers and leaves with a fine mist of water and move the arrangements to the coolest part of the room. When you open the door on the big day, the welcoming fragrance will have reached every corner.

Throughout the seasons such a glorious selection of aromatic flowers comes into bloom that it is tempting to gather armfuls of them together, like a fresh flower potpourri, and enjoy their heady fragrance all at once, lilac, mock orange blossom, hyacinths, carnations and roses tumbling over each other in perfumed profusion.

But this is one temptation that should be resisted. The widely differing scents of fresh flowers, once they are cut, do not blend together with anything like

the subtlety of a bowl of dried flower petals. When you think of it, this is not surprising, for in a potpourri the dried petals are mingled together with spices, aromatic oils and a fixative, their various perfumes linked and left to mature, like ingredients in a stew brought together with a sauce.

The perfume of each fresh flower, however, is competing with its neighbors, and it is all too easy to create a head-on clash of personalities. The result is an overpowering, cloying scent, the kind you would never want to buy in a bottle. It is better by far to limit your choice to only two fragrant flower types – roses and carnations are perfect partners, for example – and then add complementary aromatic leaves if you wish.

A MEDLEY OF PERFUMES

One of the headiest floral scents belongs to the hyacinth, which is much underrated as a cut flower. Here is one that, for perfume, is head and shoulders above its competitors when you are decorating a large room. The huge, densely packed blossoms look magnificent in a shallow bowl, the flower heads radiating outward to form a wide-topped cone. They are also useful filler material for pedestals, giving good visual weight at the base of the design.

All bulb flowers, with their fleshy stems, need a little pampering before being arranged, and are happier in water than soaked foam. Hyacinths, because of the weight ratio of flower to stem, are an extra-special case. Follow the usual procedure of cutting the stems under water and giving them a good long drink overnight. If the heads do threaten to overpower the stems, push a piece of thick wire up from the base to the tip. This will keep the stems straight or, indeed, if your design requires it, enable you to coax them into gentle, but controlled, curves.

Beautiful cut plants that smell delightful too add a charming touch to a special event. Here three white china jugs hold a loose arrangement of roses, pinks, chamomile, carnations, mint, lemon balm and wild grasses.

*Laurel leaves form an attractive, dark background for this herb ring which includes sprigs of rosemary, curry plant (*Helichrysum italicum serotinum*), lavender, sage, bay and pelargoniums.*

There is a delightfully high perfume count among flowering shrubs. With their graceful curves, branches of Mexican orange *(Choisya ternata)*, mock orange blossom (philadelphus) and lilac rise to great heights for designs on a large scale. The first two, with their gentle clusters of pure white and sweetly scented flowers, are a popular choice for bridal arrangements. They are also a good choice for groups that have to be seen literally in a poor light. Not so lilac; because these flowers are so much more densely packed they need the assistance of good, clear lighting. Without this the flower heads can "burn out" and look almost somber.

As with all woody, flowering shrubs, it is best to sacrifice the leaves at the outset, as they wilt long before the flowers fade, and include other sprays of foliage which have been specially treated according to type (more details on pages 10 and 11). Strip off the leaves, strip back a little of the bark and split or lightly crush the stem ends, and then leave them deep in tepid water overnight.

BALANCING THE FRAGRANCES

Lilies of all kinds grace any occasion with their majestic good looks and exotic scent. Whether you mingle them with other blooms or accentuate their positive outlines by featuring them alone, plan an uncrowded, open design and allow plenty of space for buds to open freely. The waxy petals are deceptively delicate and bruise easily. And, incidentally, the pollen permanently dyes anything with which it comes in contact, so be especially careful when handling the flowers.

Roses are probably the number-one favorite among fragrant flowers, and if you are planning a garden, it is worth selecting them as much for their scent as for their color. We have to go back a few generations, to the old-fashioned types, for the sweetest perfume among the damasks, Provence and moss roses.

They are such a charming feature of the garden that it seems almost a shame to cut them. To make the very most of roses indoors, strip off the thorns and lower leaves and treat the stems as for all woody material. Gather the roses into a bunch, wrap a wad of paper under the flower heads to protect them from steam, stand the stems deep in warm water, and leave them overnight. An aspirin in the water when you arrange them will help the flowers and their perfume to linger longer.

Can you think of anything more welcoming than to

open the door of a guest bedroom and sense – even before you see it – a group of sweetly scented flowers? A china teapot spilling over with pale pink and crimson old roses, a bud vase holding a few elegant stems of freesia, a cup and saucer filled to the brim with tiny garden pinks and aromatic leaves, would all extend the most gracious of welcomes.

A little flower scent goes a long way in bedrooms – where heavily perfumed blooms can be cloying – and on dining tables where strong floral fragrances compete unfavorably with any subtle blend of flavors you may serve. Not only that, they can be offputting to anyone who likes to savor the bouquet of the wine.

HERBAL ROOMS

Culinary herbs complement scented flowers of all kinds, and naturally bring their own aroma to an unscented group. A casual display of herbs – thyme, mint, chamomile, sage, parsley – all tumbling over each other just as they grow, softens the appearance of a kitchen. Don't spend too much time on the arrangement, though, as it is sure to be depleted as one stem after another finds its way into the salad.

The evergreen herbs like bay, rosemary and sage and the seedpods of others such as fennel, caraway, and dill weed are good keepers, and have their aromatic part to play in long-lasting arrangements. Long spikes of rosemary look lovely in winter with holly and ivy, and sage with its furry gray-green or purple leaves is a good "base filler" to cover holding materials. The bright yellow clusters of fennel seedpods add a hint of spice to a golden arrangement, and a "see-through" feature that is a good contrast to heavy flowers.

ENJOYING THE FRAGRANCE

Most herbs – lovage, marjoram, mint and savory, to mention only a few – are "medium keepers" and will last a few days in water. Others can be disappointing unless you single them out for special treatment. So, if you want to enjoy the sharp citrus odor and variegated colors of lemon balm, the long slender spikes of tarragon and the furry softness – to say nothing of the bright blue flowers – of borage, burn the stem ends in a candle flame before putting them in water. Scent, like color, is an intensely personal choice. My own ideal nosegay would include deep pink clove carnations, wild roses, feverfew, thyme and marjoram, purple sage, apple mint and a spray of clove-scented pelargonium leaves.

In a casually arrayed bunch of herbs and flowers that include dock, lavender, sage, borage and feverfew, variegated lemon balm and marjoram add visual interest.

CULTIVATING THE WILD LOOK

Anyone who grew up near open countryside, wilderness or the desert will always harbor warm memories of springtime wildflowers. Between the relentless encroachment of urban development and the vital protection of the Endangered Species list, however, wildflowers are no longer simply there for the taking. No matter where you live and whatever the current regulations about picking wildflowers, there is now a moral issue at stake. Wildflowers, whether subtle violets, showy orchids or rare, endangered chocolate lilies, are part of any nation's heritage, there for all to enjoy and not for just a few to plunder.

In the United States, the protection of wild plants is covered by both federal and state laws relating to specific areas. National parks, forests and recreation areas are regulated by federal laws which prohibit the picking of any plants within their confines. State laws apply to state parks, but also cover forests and recreation areas. In addition, the Endangered Species Act makes it a federal offense to interfere with wild plants that are held on this list. These plants are either endangered or threatened by extinction. The list is revised every five years and, at present, includes about 10 percent of native American plants.

In Britain and some other European countries, laws forbid the digging up of any wild plant without the specific permission of the owner of the land, and the picking of any protected plants, all those species considered in danger of extinction. Whatever the current state of the law, this is an issue on which public opinion seems to be one jump ahead. To ensure that future generations can grow up with the memories of woodlands carpeted with bluebells, meadows speckled with mauve scabious, clover and camas lilies, white marguerites, and hedges that hum with bees, we must leave well enough alone and be content to admire the plants in their natural habitat. By doing so we will be playing an important part in building up the stock of endangered species, and taking many precious plants off the endangered list.

This is not to say that we can never pick wildflowers again, and bring them indoors to assemble in delightfully casual groups. We can cut them from our own gardens and windowboxes, for one of the great bonuses of the move toward wild plant conservation is that seed and plants of native species are now widely available for home growing.

CONSERVING WILD PLANTS

Conserving wild plants is the greatest excuse there has ever been for letting a patch of your garden "go natural" and burgeon with the plants that are native to your area. For me, this means that hilly patch beyond the stream – once neatly planted out with prim annuals – becomes a haze of speckled mist in the late spring, with a mass of Queen Anne's lace drawing its pretty veil over red campion; and in the summer a gently swaying sea of wild and decorative grasses splattered with vivid scarlet poppies and sky-blue bachelor's buttons.

It is not only wild plants that are being threatened by the extension of agricultural crops, the growth of urban areas and incessant pollution. In recent years many species of butterflies, small mammals and insects have been declining too. And so if you have a garden, however small, you can play a vital part in helping to turn back this tide. By encouraging and cultivating wild plants in the garden you issue an open invitation to butterflies, moths, bees and seed-eating birds, and thereby substantially increase your own enjoyment of the garden.

Butterflies are attracted to many wild plants. In my garden their favorite is one very small patch of stinging nettles left there solely for their pleasure. They are particularly drawn to red, deep pink and orange flowers. Wild pinks and purple loosestrife are especially popular. Butterflies, like people, also favor fragrant flowers. By contrast, poppies, teasel and melilot are among the many wild plants that will have your garden humming with bees.

An autumn country harvest, which includes blackberries, wild rosehips, elderberries, crab apples and hazel leaves, joins bush roses, mahonia, honesty and Chinese lanterns in a natural display.

GROWING NATURALLY

Wherever we live, there is a long tradition of the native plants all around us being used by local people as medicines. Until comparatively recent times, natural herb medicines were all people had to calm fevers and treat ailments. The golden rule in cultivating wild seeds or buying plants is to match the plant as nearly as possible to its native conditions. Get to know your soil type – ask an avid neighboring gardener if you are in any doubt – and choose seeds that thrive on the diet you have to offer.

You may need to imitate the growing pattern of these plants as well. Most wild plants drop their seeds in autumn. According to type, some seeds germinate almost immediately, and the seedlings have to survive the winter. Other seed types do not germinate until spring, when conditions are favorable.

In the latter case you have to subject the seed (hemp agrimony with its dense masses of pink flowers and clustered bellflower are examples) to its accustomed cold spell before it will germinate successfully. This means either sowing the seeds in autumn, when in the wild they would be self-sown, or mixing the seeds with damp sand, and putting them in a plastic bag in the refrigerator for six to eight weeks. Once the seeds start to sprout, plant them out in pots in a cold greenhouse, or outside under a glass dome or an upturned jam jar. This sounds complicated, but it is not much different from sprouting beans in a drawer. The process is called stratifying.

The seeds of all leguminous species such as vetches, clovers and trefoils need a little rough treatment if they are to germinate successfully. This is called scarification, a process that breaks down the seed coat and enables the embryo to absorb moisture. All you do is rub the seed gently between sandpaper to graze, but not completely penetrate the coating. Without this assistance the hard seed coats can take several weeks to become water-permeable.

Apart from these preliminary processes – which of course are required for good germination of many cultivated seed types too – wild seeds are planted and nurtured like any other comparable plant type. Sow seeds in potting soil in trays or boxes, water them with a fine spray, cover them with glass or plastic and keep them at a germinating temperature of about 68°F. A simple heated propagating tray is ideal. Or when the time comes (read the instructions carefully) scatter the seeds thinly outside, where they are to grow – a method that seems more natural for wild seeds. Keep the seed bed evenly moist and free of weeds, and be sure to put a marker to remind you of its position. Some seed types are notoriously slow to germinate, and can surprise you months later.

ARRANGING WILDFLOWERS

If you bring a few specimens indoors, how are you to arrange them? Firstly, as you cut the stems, put them straight into a bowl of tepid water. Many wildflowers are tender and wilt easily. Study the stems carefully. The ratio of leaves to flowers is on the high side in many wild plants – breeders having "improved" on this characteristic in their cultivated counterparts. Strip off the leaves, and the stems may be transformed. Deadnettles, for example, hide their hoops of tiny red, yellow, mauve or white flowers under an awning of dull green leaves. Pick them off and you reveal rings of tiny flowers with all the charm of miniature orchids. Mallow is another whose saucerlike mauve flowers are almost swamped by foliage.

Arrange wildflowers as casually as you know how. Bunch several types loosely in your hand, pulling up one stem after another to show them to advantage, tie the stems loosely with twine and drop them into a container such as a jug, preserve jar or kitchen storage jar, or something with more panache, a silver teapot, pewter mug or pretty glass vase.

Many of those plants with slender stems are more at home in water than in holding foam, though you can ease their passage by piercing holes in the foam with a very fine toothpick.

A dainty china basket spilling over in early spring with violets, snowdrops, pulmonaria and primroses; a zingy yellow enamel pot pouring out a blaze of colorful buttercups, yellow deadnettle and sweet clover; a windowbox planted with cowslips and wild pansy; a watering can piled high with ornamental grasses and scarlet poppies – you can still create memories with wildflowers that you have grown.

Raised from seed bought from specialty suppliers, these cowslips provide a welcome reminder of open fields and country lanes.

FLOWERS IN THE OPEN AIR

Whether you have supper on the balcony on a balmy summer's evening, a cocktail party on the patio, a buffet lunch in the garden in the heat of the day, or a casual barbecue, there is always an extra "lift" to any occasion celebrated out of doors. It seems more fun, more fancy-free, more frivolous. And the ideal setting for a feast of flowers.

It isn't everyone who can boast a garden that is like a French Impressionist painting, awash with clouds of color and the perfect background for a party. Very few can, in fact. Some gardens, planned on sculptured lines, are designed to be a symphony of greens; they make good candidates for a little imported color just for the occasion. Others might go through an awkward spell, as one burst of color fades and another has yet to take over. Perhaps there is a dull corner that needs brightening, or just a patio or balcony to set the festive scene. Whatever the occasion or the setting, flowers can do much, much more than simply decorate the table.

Balconies are ideal for a little window dressing for a special party. You can weave plants in and out of the railings, loop a flower and ribbon swag over the door, or from end to end of the balcony rails, fashion flower-ball trees to stand at each end, hang shallow vases or baskets on the house wall, or arrange a simple but spectacular pedestal.

Ivy is hard to beat for trails and swags. It not only lasts well out of water – of paramount importance – but, in its various forms, can range in color from deep buttery cream through to bright lettuce green to the deepest of shades. Just a spray or two of the darkest greens for contrast is enough; in bright sunlight they look dreary, and at dusk they merge into the shadows. Other possibilities – which will naturally depend on local and seasonal availability, and whatever you can beg or borrow – include willow, honeysuckle, lime green cypress (avoid the dark shades), hawthorn, hazelnut, japonica, Solomon's seal and vine shoots.

To simulate climbing plants, simply weave the longest stems through upright supports, wiring the stems together where they overlap. Take matters into your own hands when it comes to deciding how full these plants are to be, and wire on the most accessible and long-lasting flowers around. Choose from single chrysanthemums, marigolds, marguerites, zinnias and other composites or carnations and pinks. Or take the illusion a step further and use artificial flowers.

Left: *Anemones and fuchsias appear to grow in profusion from this hanging basket. There are, however, no rooted plants. The flowers and foliage are held in a block of foam concealed in sphagnum moss.*
Right: *For a special occasion outdoors, consider decorating pillars, fences or statues with freshly cut flowers. Here a plinth bound with vine leaves and mallow supports a topknot of fall fruits and cypress stems.*

To make a swag, measure a piece of thick, coarse string, allowing enough length for it to hang in deep curves. Make up bunches of ivy and asparagus fern, and wire them to the string, each one overlapping the stems of the next. Secure the swag in place with ribbon bows and tuck in flowers at intervals, just before the party begins.

FLOWER-BALL TREES

Flower-ball trees, equally at home on a balcony or patio, can make a colorful impact in a small space. For each one, fix a sturdy 3-foot cane in a large flower pot, and surround it with soil or pebbles, topped with a scatter of soft-looking sphagnum moss. Shape some damp moss into a ball, cover it with crumpled wire netting and attach it firmly to the top of the cane. On a smaller scale, use spheres of soaked foam for tabletop trees. Push stems of trailing leaves and flowers through the wire and into the moss, creating any effect you like. You can go for broke on color and use brilliant pom-pon dahlias in reds, yellows and pinks, or introduce a hint of romance with pale or deep pink and cream roses and sweet peas, clouded over with a mist of Queen Anne's lace or baby's breath (gypsophila).

For a make-believe tree with more wayward habits, anchor a twiggy branch firmly in a pot and wire on large single flowers or mixed bouquets, or hang tiny baskets filled with fresh or dried flower nosegays.

If the table takes up all the available space, there is the wall. Put a massive pitcher of flowers in the window, attach a ribbon of flowers and leaves to the windowsill or make a hanging basket. A shoulder shopping bag lined with plastic and filled with a block of soaked foam can look spectacular. Cut the most curvaceous stems you can find – trails of Russian vine and clematis, perhaps – and pack the top with sweet Williams, ranunculus, or whatever is in season.

FINDING CONTAINERS

Moving onto the patio, you can carry these ideas with you and, with space to spare, fill tubs and pots with eye-catching, scene-setting flowers. Containers for the garden should be fairly substantial, to take account of wind velocity and party spirits. Earthenware jugs, strawberry pots and flower pots all have the right feel. If you decide to fill the neck of a jug with foam, for a downwardly mobile design, partly

Hanging on bannisters, decorating a hallway, or the walls of a small courtyard, swags lend a festive air to any occasion. For a swag that is pretty, but not overfussy, coordinate the flowers for color and shape and "sprinkle" them at random along the length.

THE COOL LOOK

When the temperature is soaring, and flowers and guests alike are feeling the heat, a watery floral scheme can be the most welcome sight. A stone sink garden, bird bath or large mixing bowl filled with water and floating with flowers, water-lily style, is pretty and practical. A collection of large glass storage jars holding cool-looking foliage and white flowers can take up a refreshing stance in a corner. Even individual carafes and old glass ginger ale bottles, each one supporting a single stem, can seem more appropriate than a crowded arrangement of any kind.

To survive the heat of the day, you need to choose the toughest of flowers, as dainty wildflowers will not stand a chance. Spray chrysanthemums, carnations, lilies and freesias are among those that keep well. As always, but particularly in the heat, it is important to cut flowers straight into water and give them a long drink in a cool place before exposing them to the sun.

If the celebrations are likely to spill over into the evening, choose flower colors that will hold their own in the gathering dusk. White, pale lemon, pink, apricot, ice blue and pale green flowers and foliage will still be drawing admiring glances long after deep reds, dark blues and purples have faded into obscurity.

Alternatively create a pool of light with white household candles in glass jars, each one partly filled with water and bobbing with small floating flower heads. Arrange ten or more colored nightlights on a flat white dish and scatter flowers among them. Bind tiny bouquets to long white candles, tie trailing ribbon bows and stick them in soil in a flower pot or straight into the ground, well out of the way of children.

fill it with water anyway. The extra weight will help to make it more stable.

Strawberry pots that are temporarily out of commission spring to life if you fill the side apertures and wedge the top with foam. A cascade of flowers and trailing leaves at the top and simple bouquets of bright garden flowers all around make a stunning display.

A collection of clay flower pots makes a cheap and cheerful feature. Place a jar of water or piece of soaked foam in each and fill them with the lightest, brightest garden flowers you can find.

Watering cans and flower baskets, wooden seed trays, a child's wheelbarrow, even a galvanized bucket – these are all containers that are naturals in an outdoor environment. Keep the feeling of wandering around the garden and cutting a few roses here, a few lilies there, or a spray of foliage. Do just that, but hide a block of foam in the basket first. Then push in the stems so that the flowers spray out in a fan shape, some extending way over the rim, others cut short and pushed close against the foam in an act of concealment. It makes a perfect decoration for a patio, to stand on a flight of steps or on top of a low stone wall.

FLOWERS ON THE TABLE

Rose petals scattered on a pink voile tablecloth for a wedding shower party, a hoop of trailing vine leaves and pretty rosebuds around crisp white damask to show off a celebration cake, a garden bouquet tucked inside the table napkin at each place setting illustrate that decorating a table with flowers can extend far beyond an arrangement in its center. Whenever there is a party on the calendar – and even when you plan a special meal for just two people – consider the table as a stage on which to set the food, and a tablecloth as a blank canvas to paint with flowers.

Scattering flower heads on the table is the charming customary way to welcome guests in North African countries. One of my most vivid mealtime memories recalls a vision of a round table set under

the night sky, china, glass, silver and stars twinkling and the cloth almost hidden beneath a tapestry of fresh flower heads. Marigolds, marguerites, zinnias, geraniums, the massed effect of dozens of separate flowers strewn on the table at random added immeasurably to the enjoyment of the food and rivaled the fragrance of the orange groves. True, in Tunisia the flowers grow abundantly and are used liberally. But the idea is so delightful that it can spark off inspiration in all directions.

The random look is pretty, quick and easy to achieve, and in no danger of getting "fussed" as the meal progresses. Match the scattered flowers to a pattern on the dinnerware. Bachelor's buttons and chamomile tossed onto a table set with blue and white china look as if they have just fluttered down from the vegetable dish; rose or carnation petals have extra appeal, as heavenly as potpourri in the making, and marigold and buttercup petals are an easy way of investing in a visually profitable crop of gold.

"Arranging" flowers out of water is not, of course, the kindest thing to do to them. So choose the flower types carefully, according to their keeping qualities, and give them the usual good, long drink in tepid water before beheading them – that is, if the decision to do so is premeditated. If it a "crime of passion" when you suddenly decide to harvest a few flowers from your garden to scatter on the table, then cross your fingers and hope they last as long as the party spirit.

Composite flowers of all kinds such as doronicum, single chrysanthemums, all the daisy types, erigeron, single and double zinnias, pyrethrums, asters (though these are rather large for this treatment) and scabious, last well. The dianthus family, all the pinks and carnations, roses in bud or partly opened bud, separated flowers from stems of delphinium, clarkia, hydrangea and snapdragons, flowers with waxlike petals such as lilies and camellia and the everlastings, helichrysum, ammobium and helipterum, should all continue to look fresh as a daisy out of water for several hours.

Get into the mood of using flower heads to create impact in this way, and you can advance from the imprecision of a random scattering and "paint" geometric or abstract shapes on a table, outlining a central feature – such as a cake stand or candlestick – or dividing the surface into place settings.

Table place settings are transformed with individual "garnishes" of fresh flowers. Given a long drink in tepid water before being cut from their stems, alstroemerias, carnations, fuchsias and anemones will keep their appearance throughout a leisurely meal.

IDEAS FOR TABLES

When a table is for a touch-me-not display, to show off a wedding cake, for example, you can simply place flower heads edge to edge, confident that they will not get jostled out of line. Ring a round cake stand with pale pink rosebuds, soft blue hydrangea florets, pastel lemon double pinks, whatever matches the color and design of the frosting. If the cake stand is square at the base, make it the center of criss-crossing lines of flowers, again echoing the style of the cake decoration.

Show off your prettiest silver candelabrum with a hoop of flowers around the base. They will double in value as they are reflected in the metal. Place a curved or straight "ribbon" of flowers and leaves down the center of a dining table, well out of the way of guests reaching for the salt, or make a hoop of tiny flowers around each coaster at the place settings.

When the table design is central to the comings and goings of the party, as a buffet table is, the flower ribbons need to be given more permanence, so divide a party table into sections with strips of flowers, the drinks at one end, then the first course and so on.

Measure from front to back of the table and cut to length either a piece of string (that will not show) or a piece of ribbon (that will). To work with the string, cut short sprays of leaves, and flowers with short stems. Use a roll of fine wire to bind the leaves and flowers, all pointing toward one end of the string, working as far as the center. Turn the string around and complete the design in reverse, the flowers pointing the opposite way. Be sure to place them at equal distance, and bind each one close enough to cover the stem of its predecessor.

Look critically at the strip and tuck in extra leaves to hide any gaps or telltale pieces of wire, and attach a special flower to neaten the parting of the ways at the center.

Variegated lamium or periwinkle leaves (condition them thoroughly first) are a pretty foil to flowers used in this way. Cut the stem just above each pair of leaves to make them look like wings on a stalk, and place the smallest ones at each end of the strip, the largest ones toward the center.

RIBBONS, SWAGS AND POSIES

Different flower types need different treatments if you are using ribbon as the base. Flat flowers with shallow calyces can be stuck onto the ribbon with modeling clay or double-sided adhesive tape. Carnations and others with long, thick calyces are best left on a short stem, made into a small bouquet with a spray of leaves and attached to the ribbon with a few holding stitches.

Depending on where you live and the season, an edge-to-edge strip of flowers can be something of an extravagance. Pad it out with delicate leaves, including only a few flowers, and the word *economy* will never enter anyone's head.

The vertical drape of a tablecloth is such a perfect "canvas" for flowers, you will never want to see an unadorned cloth again. As a pretty alternative to the draped swag, and exquisite for a summer wedding, transform a cloth with bouquets.

A large expanse of cloth, covering trestle tables for a buffet, can be covered with tiny bunches of flowers pinned or sewn at random or in rows. Use marigold and ivy leaves, cornflowers and euphorbia, rosebuds and rosemary sprigs, lawn daisies and campanula, but keep the food simple, for it will be faced with distracting competition.

For a more formal look, decorate the point of a small tablecloth with a "shower" bouquet. Make a fan shape of fernlike leaves, yellow freesia, white pinks, and yellow and white feverfew. As a finishing touch, edge the rim of the display table with a geometric line of pairs of leaves and dainty flower heads such as periwinkle and chamomile.

The rim of the table makes a natural frame to set off a buffet spread. Pin long stems of ivy or strawberry leaves around the table edge and attach a few flowers at intervals. Resist the temptation to do this for a sit-down meal. Guests might resent armfuls of foliage coming between them and the food. Flatter them instead with flowers in a personal way. Tuck a single freesia or a small bouquet into the fold of a napkin, or make a daisy chain as a floral napkin ring.

Just one cautionary note. Use stainless steel pins to avoid rust marks. But do not get carried away and pin flowers to your finest, most precious linen. The weight could tear the threads, so use lengths of inexpensive material or discarded sheets.

A "shower" bouquet of wired carnations, freesias and feverfew, finished with satin ribbon, decorates a table.

DESIGNING WITH LEAVES

Leaves are the unsung heroes of many a flower arrangement, unobtrusively providing a natural background against which flowers may be seen in all their glory. But foliage can be glorious too, and designs that focus attention on the wealth of variation in leaf color, texture and shape have an especially serene quality.

A pot of herbs on the kitchen windowsill, shiny bright green lovage contrasting with woolly purple sage, with golden marjoram, with palest cream pineapple mint, with slender chive stems, will be a restful feature in a busy area. A circlet of evergreen herbs, bay, rosemary and lavender with fennel and caraway seedpods, is a theme decoration to hang on a wall or to dangle from the ceiling. In earlier days an herb wreath would have been designed as much to keep insects at bay as for its simple good looks.

Contrast in your mind's eye these charmingly nostalgic ways to group a handful of aromatic leaves with, shall we say, a "green" pedestal of towering iris spikes, curving arcs of young beech leaves, huge fleshy misty-blue hosta leaves and ferns like complicated paper cut-outs, or a winter display of spiky rosemary, variegated ivy and dark green holly thick with scarlet berries, and the range of options begins to become apparent.

TENDING THE FOLIAGE

If it is to give long and faithful service, foliage needs special preparation and care. If you dash in from the garden with an armful of flowers and leaves and arrange them in a hurry, you will pay a price in terms of their longevity. You will find that deciduous leaves, frail and tender as they are, wilt long before almost any flowers. Only evergreens have the inner resources to withstand such shock treatment.

The immature foliage of early spring, when the sap is still rising in the plant, is most tender of all and demands the most careful preparation. Branches of yellow-green lime, so thin that shafts of light shine through them; sprays of lettuce-green oak, the scalloped leaves not yet fully grown; curving stems of beech just perfect for designs with soft, flowing lines – to enjoy the beauty of deciduous foliage, and

immature foliage most of all, you have to plan ahead and cut the leaves a day in advance to allow them time to become acclimated. Becoming acclimated simply means becoming full of water, a state known as turgid. With this built-in reserve, leaves can last well indoors, whether they are arranged directly in water or in soaked stem-holding foam. As leaves naturally draw moisture from two sources (taking it up through the branches and stems, and through their surface tissue), immersing them in water is the way to ensure the best results.

SOAKING TIME

Not everyone is prepared to put a "do not disturb" notice on the bathtub overnight while sprays of leaves are enjoying a good long soak, but in most homes there is some container that will serve the purpose. It may be a shallow photographic developing tank, an extra-large baking pan, an old drawer securely lined with a sheet of plastic, or a deep bucket. A plastic washbasin will house short branches and can, incidentally, usefully coax gentle curves – if that is what your design requires – into longer ones.

Prepare stems in the usual way (see page 11), keeping in mind that everything you do is designed to make it easier for the plant material to take up water. Strip off all the lowest leaves, as these will eventually be superfluous and either overcrowd the design or lie below the water level.

Immerse branches of mature leaves in water for several hours or overnight, and immature foliage for a maximum of two hours; any longer than this and young leaves become sodden and limp. Transfer the branches to stand upright in a container of tepid water, again for several hours if possible, or for a minimum of two to three hours.

If, after all this, the leaves start to wilt – and immature ones are the most likely to do so – you will have to resort to drastic remedial action. Bunch the

This bounty of foliage, arranged by species in concentric circles, indicates the range of colors and textures of leaves available. With a "flower" center of senecio, this display includes the leaves of scented geranium, variegated ivy, carrot, vine and maple.

branches of stems together, protect the leaves from the steam by wrapping them in newspaper, and dip the ends in boiling water for between 30 seconds and one minute. This treatment fulfills a number of functions: it breaks down air locks, sterilizes the stem ends and kills any harmful bacteria.

WATER TREATMENT

Trails of leaves on slender stems – periwinkle and Wandering Jew, for example – also respond well to total immersion. If the leaves at the tip – the most vulnerable, and always the first to go – do flag, they may be revived by burning the stem ends in a match or candle flame. If this does not do the trick, there is nothing you can do but cut off the offending tip and the wilting leaves with it, and "bury" the stem in the body of the design. Tipless stems are not elegant enough to feature at the top and sides.

Large fleshy leaves such as hosta, fig, and *Fatsia japonica* – so useful for providing visual weight at the base of the design and for concealing the holding material – have considerable water-retaining capacity and consequently have an exceptionally long vase life. Make their life easier, and your pleasure in them longer, by immersing the single leaves in a flat dish of water for several hours.

RESILIENT EVERGREENS

Evergreen leaves – yew, plain and spotted laurel, camellia, ivy, eleagnus and privet, to name a few – all last well in or out of water. Give them an initial water bath so that they can drink their fill, and they will last for a week or more fashioned into swags, drapes, ribbons, wreaths, hoops and other festive designs that have no recourse to water.

Decorating our homes at Christmastime with a swag of holly, ivy and other evergreens over a door or mantel is a tradition whose roots predate Christianity. People believed that there were magical properties in plants that could retain their leaves when other trees and shrubs had bare branches. An Advent wreath of

Subtle but not somber, this display includes aquilegia, purple sage, honesty, totinus, berberis, maple leaves and love-in-a-mist seedpods.

cypress, juniper and yew linking the four symbolic candles, a ribbon of soft trails of ivy and honeysuckle draped in front of a buffet table, a hoop of lavender leaves and flowers to perfume a clothes cupboard, are some of the countless ways of using evergreens.

Remember that leaves are quite capable of retaining water on their outer surfaces. So to prevent water from dripping onto furniture and other precious surfaces, toss sprays of wet leaves gently over and over on newspaper or a piece of cloth, and pat dry large leaves – such as laurel – with papertowels. Any evergreens that have been covered in city grime and do not emerge from their water bath sparkling bright will benefit from being swished in warm, soapy water before being rinsed and dried.

LEAVES AS HIGHLIGHTS

In all its wonderful variety, foliage deserves to capture the spotlight occasionally. Try mingling shiny leaves with matte and woolly ones, plain leaves with speckled, striped or otherwise variegated ones, and colors of every tint, shade and hue from palest cream to deepest purple. You can choose to group a subtle blend of gray, green and white foliage, capture the glory of falling leaves in the bronze, red and yellow range, pick a sharp scheme in lemon and lime colors, or create a medley including all these ideas. Do not let your selection range over too wide an area, though. A display of four or possibly five species at the most will enable each to be enjoyed in detail. More would be muddling.

Take the shape and form of leaves strongly into account – tall, slender spikes for pointed outlines, soft trails for curves, large "flat" leaves like ivy and begonia for weight at the base. And for the focus of attention, occupying that hallowed spot at the base of the stem? Round materials are easy on the eye. You can use a single succulent, or a group of three; a cluster of round leaves – some pelargoniums are like pretty multicolored rosettes – or end-of-stem sprays of magnolia or camellia are especially charming (if somewhat wasteful) at the bud stage.

There are many different kinds of seedpods, grasses, bracts, fruit and vegetables to use too. Long spires of delphinium seedpods, furry little lupine pods, the stately "urns" that carry poppy seeds, clusters of lime-green or purple-pink euphorbia and elegant stems of heavily veined Bells of Ireland bracts, rosy-green apples and pears, limes and bell peppers – when you "go for green," there is no shortage of choice.

FASHION AND FLOWERS

Tuck a handful of flowers in your hatband just because it is a lovely sunny day or make a daisy chain for a child to wear. Pin a boutonniere to your lapel on your way to a wedding, or present a bouquet to a visiting speaker, and you are carrying on traditions many of whose origins are lost in the mists of time, since the use of flowers for personal adornment goes way back in history.

The choice of flowers and leaves was sometimes dictated by religious belief, often by custom. In ancient Egypt garlands were bestowed on rulers and high achievers. They were brightly colored, and fashioned with mallow and poppies, or made of palm and papyrus leaves entwined with herbs and the sacred flower, the lotus.

The Greeks and Romans had a strong feeling, too, for wreaths, garlands and chaplets that, with a highly practical streak, they made mainly from evergreen leaves – laurel, bay, box, ivy and yew – adorned with oak leaves, acorns, fruits, and both "sculptured" and scented flowers. Tulips, hyacinth, honeysuckle and lilies were favorites and also among those flowers that last particularly well out of water. These garlands were worn around the head, shoulders or neck and were also carried ceremonially high on staves. Flowers and petals were strewn everywhere, on furniture, floors, and over honored guests and were maybe the origin of the custom of showering brides with rose petals and confetti or rice.

MAY DAY CUSTOMS

It was the Romans who introduced the delightful custom of holding festivals of flowers. The most notable of these, dedicated to Flora, the goddess of spring, survives to this day in May Day celebrations. In some country villages in England children wearing flowers in their hair dance around a May Pole, interweaving colored ribbons as they go. The highlight of the event is when the queen of the May – usually a small child – is crowned with flowers. In Greece on May Day you can still see elderly ladies sitting outside their houses with long necklaces of fresh flowers vividly outlined against their stark black dresses. And on every door on the first day of May

there hangs a stefani, a hoop of marguerites, marigolds, geraniums, snapdragons, scabious, or whatever the garden or countryside grows.

In the city of London, in the 13th century, priests at St. Paul's Cathedral wore floral crowns on their heads, and throughout the Middle Ages judges carried bunches of herbs and fragrant flowers to sweeten the path of their procession. These delightfully named "tussie-mussies" were composed of herbs and aromatic flowers, with a view to their qualities as both disinfectant and air freshener. For old times' sake, they are carried still by English judges in session, which proves that wearing and carrying flowers is by no means a female prerogative.

The 19th century raised the fashioning of bouquets to a fine art, leaving us a charming legacy – a formal design of rings of wired flowers encircled with a protective layer of leaves or a lace doily. Ladies in that era wore ornate sprays of flowers in their hair and on their dresses. The boutonnieres worn by the gentlemen were scarcely less showy, nosegays of mixed flowers and greenery verging on the exuberant.

Whether bouquets were carried or worn, in these times there was likely to be more to the choice of flowers and leaves than met the eye. For this was the age of the hidden message, and bouquets given and received would speak volumes of love, sorrow, joy, anxiety, in a language of flowers handed down from one generation to the next.

With so much tradition to draw upon, and just for the joy of it, let flowers go to your head, and your clothes. We have ideas, some conventional, some imaginative, to circle summery hats, with roses, wildflowers and grasses; these are delightful variations for a garden party or outdoor wedding. We become more demure with neat bouquets to carry on formal occasions and indulge in some mild flights of fancy in jewelry and boutonnieres to match every

Garlands made of flowers, such as spray chrysanthemums, are a refreshing way to enliven a dress. This thick wire hoop was covered first with willow leaves.

mood. And, to make sure the flowers live up to expectations, there are detailed instructions on how to wire and attach them.

HEADGEAR

There's a really special feeling about wearing flowers in your hair and on your hat. It is quite impossible to be your ordinary, everyday self with a red rosebud tucked behind your ear, or a bunch of primroses clipped to a hairband, celebrating the joy of early spring.

What child does not feel excitement the moment there is a daisy-chain crown or a wreath of pink and blue cornflowers to wear? Who does not feel an extra hint of elegance and refinement as they pin a flower spray to a special hat to go to an outdoor party, or experience a bit of midsummer madness as they circle a favorite straw hat with a casual hoop of flowers? There are so many ways of expressing the mood of the moment with flowers that go to your head.

It seems the most natural thing in the world for children to link the stalks of lawn daisies, splitting each stem close to the flower and pushing the next one through it. And at that age, no one tells you to stand the flowers in water first, to give them a refreshing drink. You can take the idea a step further, and make a simple chain in a similar way, linking marigolds, marguerites, single spray chrysanthemums and other daisylike blooms. Make a complete circle for the child to wear as a crown, or a strip to wear as a headband. In either case, a couple of hairpins will keep it in place.

To give the decoration a little more substance – as you would for a flower girl's headdress – make up small bunches of flowers and bind them with fine wire or twine to a headband that has been purchased or, for a crown, to a circle of thick cord. Arrange the bunches so that each one conceals the stalks of the next, take a critical look at the nearly finished design and tuck in extra flowers or leaves to fill in any gaps that may be left.

Teenagers, and anyone who is young at heart, might prefer a floral headdress with a wayward look. In the French countryside at festival time, young girls love to wear crowns of braided grasses, thick bunches of stalks twisted to make a circle, the long seedpods radiating out unevenly, like shafts from a pinwheel. These crowns are worn on the back of the head and have colorful flowers – larkspur, clarkia, cosmos, mallow, sweet peas, Californian poppies – tucked in among the wild, wayward stems.

Grasses in all their delicate outlines and subtle colors, from sun-bleached cream to deep purple, are a happy choice for decorating a summer hat. It's the kind of thing to wear on a leisurely day in the garden or during a picnic by the river. Choose a straw hat with a deep crown and wide brim. Measure and cut a piece of thick string to encircle the crown and join the ends. Make thick bunches of mixed grasses, 15 to 20 stems in each, cut the stems short and bind them with fine wire. Bind the bunches to the string so that the heads will flop out, at random, over the brim, and slip the grassy hoop over the hat.

ALL AROUND YOUR HAT

There is an air of informality about an all-around-the-crown floral decoration. Substitute tiny bunches of buttercups and daisies, candytuft and love-in-a-mist, geum and chamomile – what you have, and what you please – for the grasses, and the hat will be perfect at a summer party, a tea party on the lawn or a country wedding.

For more formal occasions, you might prefer to emulate milliners' designs, and arrange a cluster of flowers at the back of a hat, color-matched to a bow or ribbon band. Roses are a perfect choice. Cut them well in advance, crush or split the stems (see page 11) and leave them in a cool place in water. Make it your last task before leaving, to sew or pin the flowers to the hatband.

If you have a long trip by car, do not subject your flowery headgear to a hot and bothered ride on the back seat. It might seem somewhat excessive, but an actress who is much in demand for garden parties

This rural headdress with a base of oats and wild grasses, includes godetia, sage, cosmos, mallow, montbretia, and fragrant garden pinks.

always travels with her latest flower creation cool as a cucumber in an insulated plastic box – a thoroughly modern version of those elegant 19th-century hatboxes.

VICTORIAN BOUQUETS

There was nothing new, in Victorian times, about the custom of bouquets of leaves and flowers being offered by shy young men as love tokens to demure young ladies. For centuries that had been an accepted way of declaring affection. But, in keeping with the times, a new formality evolved in the design.

The bouquets became completely stylized, with

Deep salmon-colored rosebuds and pinks, in a spherical bed of primrose and pelargonium leaves, make a delightful wedding bouquet.

choose compact flowers that can easily be arranged in well-defined rows. Each flower or floret is separately wired so that the handle becomes a slender bundle of wires, each one individually bound with florist's tape and then with concealing ribbons.

Rosebuds are a natural choice, their tightly curled petals making perfect circles. (Remember, though, that the buds will gradually open even when the flowers are not in water; so allow them a little space to do so.) Small double spray carnations, love-in-a-mist, cornflowers and tight, double narcissus are among the most suitable of the "round" flowers.

By contrast, single heads of freesia, delphinium, alstroemeria, flowering tobacco, campanula and sweet pea, with their trumpet and fan shapes, give variety, and help to lighten the design.

Leaves have an important part to play, providing natural contrast to the flowers and forming a protective ring around the outside – sometimes reinforced, in the old-fashioned tradition, by a stiff lace paper doily. Your choice of materials, whether they be primrose and violet leaves, geranium and nasturtium, ivy and camellia or clematis and geranium, will significantly affect the appearance of the design.

Bridal and presentation bouquets are destined to be in the limelight for several hours, so the flowers and foliage need meticulous preparation if they are to last the duration. Give all the plant material a long drink, overnight or for several hours, in tepid water in a cool room. Handle the flowers as gently and carefully as possible to avoid bruising, and spray the finished design with cool water. Stand the bouquet upright in a jar or vase and keep it in a cool place – even the refrigerator – until the big moment arrives.

WEDDING DAY FLOWERS

There is no occasion to compare with a wedding for an exuberant display of flowers to carry and wear. Weddings provide many opportunities for creative flowercrafts whether you decide to make bridal bouquets, headdresses, clusters of flowers on dainty satin slippers, or a cascade of blossoms trailing from a prayer book, or outlining a lucky horseshoe; extravagant corsages and flower-encircled hats for the female guests, or traditional or imaginative boutonnieres for the men.

If you are entrusted with the task of designing the flower arrangement to be carried by the bride and flower girls, the style of the outfits, the color and the texture and weight of the fabrics will have a bearing

rings and rings of flowers and leaves arranged, like ripples in a pool, around a central flower, with never a petal out of place. Often the central flower, slightly raised above the surrounding rows, would be a single perfect rosebud, the ultimate symbol in the language of flowers, for it signified the blossoming of new, young love.

With centuries of romantic symbolism behind them, these Victorian-style bouquets have been popular ever since for young brides and flower girls. With their neat, compact and easy-to-carry shape, they are also perfect presentation gifts for visiting speakers and guests of honor.

To achieve the tightness of the design, it is best to

on the flowers you choose for both bride and groom.

Sketches are a great help. Whether the dresses are being chosen from a wedding catalog, made from a pattern or designed by a dressmaker, ask for a sketch and make tracings of the designs. Draw outline shapes of bouquets on the outlines of the dresses. You will be able to see whether the designs you had in mind are suitable for the dresses. At this drawing-board stage it is easier to modify ideas, pool suggestions and come up with finished designs.

If the dresses are being made up, take a cutting of the material to a flower shop and note down ideas for flower textures and colors that will complement the materials, but not pale into obscurity when held against them. Buy a few stems of flowers for the bride to see. Hold them against the dress materials and check that the balance is right. Heavy flowers with waxlike petals, such as tulips and lilies, may look overpowering against a fine voile or dainty net; tiny flowers such as lilies-of-the-valley and snowdrops could well be lost against a self-embroidered material or intricate lace.

These sample flowers will not be wasted. Have a trial run and practice wiring them until the specific technique for each type becomes familiar and automatic to you.

If the bride's choice is for a perfect match of flowers and fabric – cream roses with cream silk, shell pink camellias with shell pink cotton – design bouquets with a delicate frame of leaves or contrasting flowers that will allow the main flowers to be clearly appreciated.

These guidelines apply equally when you are

Above: *Break the conventional boutonniere habit with a purple iris, a cream orchid or a crimson rosebud.*
Right: *Imaginative flower designs for a wedding.*
Below: *Wire small bunches of flowers before attaching them to the headdress base.*

A simple summer dress is transformed with a bunch of wild roses at the waist and a ribbon of flower heads around the hem. Flowers on the hat are first wired to a strip of ribbon.

designing headdresses, only now you have to consider not only the garments, but hair coloring as well. A delicate outline of leaves surrounding a flower headband or crown may be all you need to ensure that cream gardenias or bronze fuchsias are equally flattering to both blonde and brunette tresses.

However casual the finished effect is to be, the preparation of the flowers must be meticulous, so that they will not only look delightful but give a feeling of complete security. This is not the occasion to twist a few wildflower bouquets onto a stem of ivy and call it a headdress. In almost every bridal design – or indeed for any important occasion – the flowers should be wired individually and then bound into a bouquet; onto a headband or wire ring for a crown; or grouped together into a dress spray or corsage.

Wiring flowers is surprisingly quick. Each flower head or floret is cut from its stalk and mounted onto a false stem or wire – which is much more malleable than any natural stalk. You can angle the flowers in any way you wish, twisting the wire so that the flowers face upward, downward or to either side. In addition, the wire stems – thinner and less clumsy to hold than a bunch of woody stalks – can themselves be shaped to give the right-angled handle grip essential for a cascade bouquet.

It may seem less important to wire flowers for a dress spray or corsage. A cluster of wild roses and a few stems of lily-of-the-valley, or a couple of stems of freesia and a rosebud, can easily be arranged, but even so the bulk of the stems may be a problem. A slender line of wires bound with narrow ribbon or florist's tape looks more elegant, and is more comfortable to wear.

SUMMERTIME FASHIONS

Memorable occasions such as a summer festival parade, a display of country dancing at an annual fair, a cocktail party on the lawn or a lazy trip in a small boat are just made for the imaginative, and even slightly impractical marriage of fresh flowers and informal fashions.

Flowers have the ability to create the prettiest impact in the shortest possible time. A scattering of single flowers on a swirling skirt, sprays, bouquets, flower chains and flower-covered ribbons all capture the mood of the moment and are completed in a matter of moments, too.

Enhance a favorite summer dress by sewing or sticking flowers to the shoulder straps and around a ruffle or the hem; or scatter separate flowers across a skirt or over the bodice. Large stitches through the flower centers and fabric will hold them in place, and they can be pulled out easily when the festivities are over. Double-sided sticky tape can also be used.

Flowers such as marigolds, ox-eye daisies, small single spray chrysanthemums and ranunculus are suitable, since they last well out of water (after a good long drink), and since they are flat, they do not inhibit movement.

Ribbons of flowers can be worn in so many ways: sew or stick flower heads such as scarlet impatiens onto a white sash to completely encircle the waist like a waistband; or attach a few flower heads randomly, here and there along the trailing ends of a ribbon bow. Alternatively, attach small mixed bunches to the ends of a narrow ribbon belt. All these ideas will transform a plain dress in minutes.

In Spain and Portugal boys and girls in festival parades wear their floral sashes diagonally, over one shoulder and dipping below the waist on the other side, a style that derives from the baldric, a heavily ornamented sash worn to support a sword or bugle. Flowers are much more fun. Young children love to be decked out in this way, and the brighter the flowers, the better. Yellow and orange poppies on a white ribbon, maybe, or many-colored cornflowers on pale blue enter into the fun of the fiesta. A matching wreath in a small girl's hair, or a bunch pinned to a hair pin completes an outfit.

FLOWER JEWELRY

Flower jewelry is by no means the prerogative of the very young, but perhaps just the young at heart. In country districts in Greece, for example, you can see women, when they gather together to celebrate festivals and weddings, heavily festooned with brilliant garden and wildflowers. Small girls may wear bracelets and anklets of the brightest flowers they can find, such as geranium or delphinium flowers picked from the stem and threaded onto thick thread. Older girls may strive for more formality and elegance, and thread flowers to make a choker-style necklace or a rope of flowers, hanging almost to waist level.

Perhaps the pride of the family belongs to the oldest members. Grandmothers dress in black blouses, skirts and cardigans, and, unselfconsciously, wear the biggest, brightest flower jewelry of all, a thick rope of sunshine yellow doronicum or marigold flowers reaching almost to hip level.

When there is no time, and no need, for the more formal wiring techniques described on page 61, use the quick and easy threadneedle system. Gather together a generous pile of identical flowerheads, bright pink mallow, yellow *Kerria japonica*, golden gazania, or whatever grows in abundance and matches your outfit. Use a thick darning needle and fine twine, and thread the flowers to make a brightly colored, tightly packed floral rope. A long hoop of flowers that can be slipped easily over the head can be made continuous, with no clasp or other fastening. Shorter necklaces can be threaded onto the twine with long ends left to tie into a bow.

Chapter 2
DRIED FLOWER-CRAFTS

If you have ever walked around a beautiful garden and wondered whether it was possible to capture the shapes, colors, even the fragrances forever, there's good news: you can. Through a variety of simple processes you can build up a delightful collection of flowers, leaves and seedpods throughout the year, and bring them together in almost everlasting arrangements.

A basket filled to overflowing with dried midnight-blue delphinium, silvery honesty "moons," orange Chinese lanterns and yellow-eyed narcissus for a pedestal table; a romantic bouquet of palest pink and cream rosebuds, tiny deep red zinnias and creamy white chamomile; a miniature bouquet of buttercups and daisies to pin to a showy hat; a wall-hung swag that echoes the end-of-summer tints – with fluffy golden-rod, zingy yellow tansy heads, glycerin-preserved beech and oak leaves, and chestnut brown seedpods; or a glowing bowl of winter berries – all these designs and many more you can create by preserving plant materials.

Think of summer flowers as reserves in the bank. You harvest and preserve them when the gardens, hedges and markets are a blaze of color and supplies are plentiful. You store them carefully and draw on them, when the gardens are empty and market supplies expensive.

THE BASIC TECHNIQUES

There are four basic preserving techniques. The first, air drying, is suitable for all types of flowers that are formed of dense clusters – such as gypsophila, tansy, sorrel – and for multipetaled composites.

Double zinnias and double marigolds are examples. This is also the method for preserving a whole range of seedpods, from towering hollyhock to tiny rue. There are detailed instructions and descriptions for achieving this "second harvest" on pages 80-82.

Another preserving technique, which captures and holds the shape and form of single, open flowers like buttercups and pansies, and trumpet shapes like narcissus, snapdragons and orchids, involves using a desiccant. In this method a dry powder is sprinkled gently around the flower; gradually, and without altering shape or color, the desiccant – and several different types are available – draws out the moisture.

The art of pressing flowers and leaves is a favorite childhood hobby; it seems like magic to be able to close beautiful shapes between the pages of an album and then rediscover them, transformed but still retaining their recognizable outlines and colors. Pressed materials have many decorative possibilities – we explore these in later pages.

Preserving in a solution of water and glycerin is the fourth main technique we will be examining in detail. This produces rich, strong colors – sprays of glowing brown beech leaves, supple lime leaves complete with their eliptical fruits, prickly stems of glistening blackberries, these and many other preserved materials will add immeasurably to your designs of fresh, dried and even artificial flowers.

Above: *Dried rosebuds make a pleasing display in a china basket.*

Opposite: *There is a wealth of plant material that can be successfully dried and that will enable you to fill your home with the fragrances and colors of the garden.*

PRESERVING TECHNIQUES

The simplest of all preserving techniques is air drying. According to type, flowers, seedpods, grasses and some leaves may be dried in a warm, airy place, hanging in bunches, standing in a container or spread out on a flat surface.

AIR DRYING

In the process of drying in this nearly natural way, flower petals shrivel to become more dense. This shrinkage is almost imperceptible in flowers that are composed of a high number of petals.

The conditions under which you harvest plant materials for drying play an important part in determining the success or failure of the process. Choose a dry day, and as far as possible the driest time of day. This means waiting until the dew has evaporated in the morning, and gathering the materials before the dampness settles again in the evening. It is almost impossible to achieve a high standard, drawing out the natural moisture within the plants, if you have to contend with atmospheric moisture as well. Excess dampness is almost inevitably a recipe for mold.

The stage of development of the plant material is another critical factor. Most flowers and seedpods for drying should be cut just before they are about to open. If you allow thistles, bulrushes and pampas grass, for example, to open fully, they will explode since this is their natural process of seed distribution, and the moment for drying them will be lost. Long flower spikes should be cut when the topmost third is still in tight bud.

Next, the time scale. Plan to harvest the materials on a day when you will be able to prepare them for drying with the shortest possible delay. Put them in water, as for all fresh materials, as soon as you cut them.

The temperature and humidity of the "drying room" also has a significant effect on successful and realistic results. Dampness and strong sunlight are the extreme conditions to avoid. Choose a well-ventilated room that, if it cannot be left with closed blinds, at least has dark corners. A garage is suitable only if it is well insulated and does not get steamy in a rainstorm, a kitchen only if it never suffers from condensation.

The perfect spot will vary from one home to the next. I dry all my flowers in one end of the room we use as an office – there is no excuse for lack of inspiration, sitting looking at a false ceiling of bunches of astilbe, senecio, cornflowers and helichrysum. It could be that a spare bedroom, the corner of a landing, a stairwell, a sewing room or utility room would be just the place. Although the plant materials are best left undisturbed while they are drying, there is no need to hide them away; hanging in bunches or standing in pots they make an interesting and decorative feature.

The largest candidates for air drying – long spikes of delphinium, mullein and purple loosestrife – and tightly closed rosebuds benefit from drying more rapidly at a higher temperature. A warm linen closet or the space above the boiler would be suitable.

BUNCHED OR FLAT

From the list on pages 76-78, you will see that by far the greatest number of plant types may be air-dried hanging in bunches. Another group, those that have tiny, delicate flowers, like gypsophila, all the

An empty shelf or drawer, if well ventilated, can provide a convenient place for drying plants.

umbrella-shaped heads, like yarrow, fennel and lovage, and dainty grasses are best dried upright, their stems stuck into dry foam or standing in roomy containers. A convenient way to dry the spherical flower heads of the onion family – from the diminutive chives to the showy *Allium aflatunense* – is to push the stalks through the grid of a wire cooling tray suspended between two supports.

Although it sounds like a contradiction in terms, some flowers are dried most successfully by being left standing in water. Hydrangea flowers in all the heavenly shades of blue, pink and green, small peony heads, white chamomile, heaths and heathers, and pom-pon dahlias keep their colors and characteristics best if they are left for two or three weeks with their stems standing in about ½-1 inch of water. As the water evaporates, so does the moisture in the petals. Once the flowers have dried, wash and then thoroughly dry the stems before storing them; otherwise any residual dampness will spread.

Some grasses and other seedpods supported on slender stems are best dried flat. You can use trays, box lids, sheets of newspaper, wrapping paper or open shelving, but not a closed drawer because the process depends on a free circulation of air. Turn the materials over occasionally while they are drying, so that each surface in turn lies uppermost.

Pliable stems such as broom, willow and mulberry can be twisted into a circle, tied firmly and hung on a hook to dry. When they are untied they will spring into the kind of gentle curves that give natural "movement" to a graceful arrangement. The circular hoop of stems, just as it is dried, makes an unusual and natural alternative to a florist's wire or foam ring as a base upon which to attach herbs or dried flowers for a wall hanging.

Very few leaves preserve their characteristics when air-dried. Carrot and bay, which can both be dried hanging in bunches, are exceptions. Others, as will be seen, are best preserved by pressing or in glycerin solution.

When preparing stems for drying, strip off all the leaves, removing them carefully if you wish to treat them by another method. For hang-drying, gather the stems into bunches and tie them firmly with a slip knot that can be tightened as the stems shrink. It is better to make several small bunches rather than one overcrowded one, so that all the materials are adequately and evenly exposed to the air. Fan out the stems to dry in containers so that the heads are not touching each other, and arrange those to be dried flat in neat rows in a single layer.

The drying time depends on the size and density of the plant material and the temperature and humidity of its environment. When they are completely dried, flowers feel and sound crisp and papery, and seedpods rattle.

Once plant materials have dried, you have several storage options. If they have been dried in overheated conditions such as a linen closet, you must move them to normal room temperature; otherwise they will become too brittle to handle. But if you have come to appreciate the decorative properties of drying flowers around the house, you can leave them where they are – they still need a dry, warm and reasonably dark place – or store them in boxes.

USING DESICCANTS

Legend has it that centuries ago a lady discovered by happy chance the delightful art of powder-drying flowers. The story goes that she buried tiny, fresh red rosebuds in crushed sugar, hoping to create a perfumed sweetener for a special confection. And when she extracted the flowers she discovered that she had not only achieved her culinary aim, but had preserved the buds perfectly, their charming form and color intact.

Drying flowers in a desiccant of some kind – crystals or powder that gradually draws out their natural moisture – greatly extends the decorative range of your dried flower collection. Many flower types, as we have seen, may be dried simply in a free circulation of warm, dry air. Many others – some large trumpet shapes, some minute specimens – can only be preserved successfully if they are completely immersed in a drying medium.

Just listing a few of the flowers that can be preserved in this way brings to mind delightful little bouquets to arrange for a dressing table, small individual groups composed in wine glasses, to decorate each place at the table or tiny baskets for small flower girls to carry, or to take as a long-lasting gift to someone who is ill or has no garden. Rosebuds, sweet peas, garden pinks, double night-scented stock, violets, buttercups, daisies, side shoots or separate florets of hydrangea and larkspur, lilac, hyacinth, broom and lily-of-the-valley – these will be enough to keep you busy throughout the seasons.

When preparing to dry flowers by this method, as by any other, it is important to harvest them when they are absolutely dry. If you attempt to process them in desiccant when there is residual moisture clinging to the petals or lurking deep inside a cone shape, the raindrops or dew will dampen the drying agent and render it ineffective before it starts to draw water.

The principle is simple. You make a layer of desiccant in a plastic box or canister and place the flowers well apart so that they are not touching each

other. Fill deep specimens such as orchids and snapdragons with the drying agent, and then sprinkle desiccant gently and carefully over and around them. Another layer of the crystals or powder on top of the flowers absorbs the moisture as it rises. You then cover the box with a lid, seal it all around with tape and place it where it will not be knocked.

DRYING AGENTS

There are a number of suitable drying agents and, if you are impatient to get started, you may well find you have at least one in the house. Silica gel water-absorbent crystals – the kind that are used to prevent rust in cameras and clocks – are the most effective. Large crystals are too weighty for all but the largest double flowers, so buy them already ground to very fine crystals, about the consistency of superfine sugar, or process them yourself in a blender or food processor. As it is vital to the success of the operation that the desiccant is itself absolutely dry, silica gel offers a high degree of reassurance. When they are dry, the crystals are bright blue. As they absorb moisture they turn pale blue and eventually they become pink – by

which time they need to be reactivated.

Reactivate crystals by spreading them on a cookie sheet and drying them in a cool oven. Alternatively, tip them into a skillet and shake them over a low heat, until they turn bright blue again. Strain the crystals to remove any petal particles and store them in an airtight canister for reuse.

As long as you dry it thoroughly and strain out any stray matter, you can use any drying agent over and over again, as you will certainly want to do at the height of the season.

Sand, which is considerably cheaper than silica gel, is a good choice when you want to dry very large flowers or to work with large batches. White silver sand, which you can buy from nurseries, is the easiest to use. It has small, free-running particles, and is usually clean, dry and ready to use. This is not the case with other types of sand. Some sands must be washed several times in buckets of clean water until the water remains clear. These sands can be drained, spread on cookie sheets and dried in a cool oven. Leave them to cool thoroughly before using. Any warmth in the desiccant could scorch the petals.

Other water-absorbing agents are domestic borax, salt, powdered starch, yellow cornmeal, alum powder, white detergent and – returning to our innovator – superfine sugar. Within this general context, these agents have varying properties. Borax and starch must be first sifted through a fine strainer, as they tend to form lumps when even slightly damp. Both are a little difficult to remove from dried petals. Sugar may prove too heavy for use on delicate flowers, and detergent, with its slight stickiness, may be difficult to remove from flowers with complex shapes.

The ideal solution often lies with a mixture of two desiccants. Tried and tested "recipes" include: equal parts of ground silica gel crystals and sand; one part sand to two parts borax; three parts sand or alum to one part borax; two parts corn meal to one part borax;

For an eye-catching presentation place potpourri in an elegant bowl and scatter rosebuds on top.

and three parts borax to one part table salt. Bear in mind that the smaller and more delicate the flowers are, the more the weight and stickiness of the desiccant may become a problem. But you can mix and match the drying agents to suit your own selection of flowers, and your own experiments.

While flowers dry perfectly in desiccants, most stems, apart from woody ones, lose so much moisture that they collapse. Added to this, you need deep containers indeed if you are to dry most flowers on a usable length of stem. Cut stems supporting a single flower to about 1 inch long.

Push the stems vertically into the layer of desiccant in the container so that the flowers are upright and spaced well apart. Then trickle the desiccant slowly through your fingers so that it fills every crevice in the flowers and touches every surface of every petal. Flowers with a deep "cup" are easier to fill if you support them in one hand, sprinkle in the desiccant and then gently lower them into the box and cover them with more of the drying agent.

This is the critical stage. If you let the desiccant fall too heavily and quickly onto the flowers, it will crush them. And if they are misshapen at this point, they will be misshapen when they are dried.

Flowers that are formed in clusters along a stem, such as lily-of-the-valley, hyacinth and larkspur, are treated in a slightly different way. Whatever shape of container you use, hold the stem so that the flower bells are inverted and fill the flowers with desiccant. If you are using an upright container – such as a coffee can – push the stem into the layer of desiccant

covering the base and sprinkle in more drying agent to build it up around the stem. If you are using a shallow box, gently tilt the container and the stem, and lay the stem flat on the desiccant layer without tipping out the contents of the flowers. Cover the flowers completely, then cover and seal the container in the usual way.

STORING AND SUPPORTING

The drying process takes from about two days (for buttercups, daisies and pansies) to five days (for pom-

Fine, white sand that has been washed and dried can be used as an efficient desiccant to draw the moisture from multipetaled flowers.

pon dahlias and double African marigolds). Test the flowers as soon as you think they should be ready. Gently brush aside the top layer of desiccant, and if the petals sound papery, remove the flowers from the box. Once they are dry, they should not be left any longer as they will become too brittle.

Use a small, fine camel's-hair paint brush to flick away any crystals or powder clinging to the petals. Push a stem wire (you can buy them from florists) up through the short stem and out through the flower head. Make a short, thin hook in the top and draw it carefully back through the flower to secure it. Long flower spikes such as hyacinths simply need a wire pushed through the stem. Store the dried flowers with their false stems stuck into dry florist's foam, taking care that they do not crush each other.

You now have the makings of a varied collection of realistic-looking dried flowers on the most unrealistic-looking stems. This imbalance is easily rectified. Bind the wire stems neatly and tightly with a thin strip of green or brown crepe paper, or with florist's tape.

The stems probably will not show. Dried flowers look prettiest when they are arranged close together, but not overcrowded, the stems supported in dry brown florist's foam or crumpled wire mesh.

A silver bon-bon dish packed with tiny cream rosebuds, sprays of golden broom and delphinium flowers; a glass cake stand piled high with a pyramid of camellias and Mexican orange blossom, a Victorian cup and saucer overflowing with godetia in every shade of pink – that imaginative and experimental cook started something with her perfumed sugar.

USING YOUR COLLECTION

As your collection of dried flowers grows – with, for instance, lavender hanging from coat hooks and coathangers, pots of water-dried hydrangeas and foam blocks that have been pierced with myriad wire stems of powder-dried blooms – the time will come to create lovely, long-lasting designs for your home and for gifts.

The wonderful thing about dried flowers is that you can arrange them in or on any kind of container, from a priceless vase to a wooden spoon, since water retention is not a consideration.

Dried flowers of all types seem to have a will of their own because they are so light. Place a few stems in a narrow vase, turn your back, and they will have popped out or shot upward. And so you need to take a firm hand and employ a few tricks of the trade.

Designs created with bunches of dried flowers rather than single ones look fabulous; dried bachelor's buttons or chamomile are, for example, more stunning if massed rather than arranged so that each stem is placed separately. In addition, the bunches are a great deal easier to handle.

You probably will not need any holding material if you arrange bunches in wide-necked containers such as earthenware pots, deep shopping baskets or copper preserving pans. Wide, shallow containers can be a problem. If the bunches will not stay anchored, fill the container with a piece of crumpled wire mesh, twist a wire around each flower bunch and hook it over the netting.

Crumpled netting in the neck of narrow containers tames dried stems just as it does fresh ones. Dried stems or false stems of wire are much more slender than fresh ones, however, and so will need a base of netting that has smaller spaces.

There is a special stem-holding foam for use with dried flowers. It is usually brown, has a tight, compact texture and takes a much firmer grip on stems than does the standard green type. Use the brown foam and you can create designs with dried materials akin to all those described for fresh flowers, with the added advantage that, with no maintenance at all, they will give lasting pleasure.

Whether you decide to fashion an equilateral triangle design in a classic urn or an armful-of-flowers effect with foam wedged into the neck of a pottery jug, maximize the impact of the dried flowers and avoid a "spotty" look by grouping several of each small type together. A cluster of pink or white helipterum, a handful of chamomile, a clump of lavender or sea lavender makes a stronger impact. Take a critical look at dried flower arrangements of all kinds and carefully fill in any gaps that reveal spindly stems. Generous profusion is the effect you are after.

Foliage does not feature prominently in dried flower design, for of course very few leaves can be dried successfully. Small sprays of foliage preserved in glycerin provide a hint of sheen in a group of flowers where the charm is in their matte paperiness. Take care not to overdo it, though: the deep, strong shades of preserved leaves can all too readily become dominant.

Other dried materials can perform the "neutral background" function normally required of foliage. Clusters of lime green alchemilla mollis, pale green bracts of Bells of Ireland, silvery honesty "moons," parchment-colored seedpods and delicate grasses all complement the modesty of dried flowers.

FOAM TECHNIQUES

When a design will give joy for months to come, it is worth spending a little time in its creation. By using foam shapes, you can fashion dried flower cones, trees, circlets and swags. Place a pyramid of foam on a cake stand, cut short the stems of dried flowers such as helichrysum, rosebuds and poppy seedpods, and arrange them in a pattern. Anchor a sphere of foam to the top of a cane, "plant" it in a flower pot and make concentric rings of dried flowers around it. Buy a plastic bouquet ring pre-filled with foam, and fill it to capacity with your prettiest flowers. Short snippings of goldenrod, clarkia or larkspur tucked in break up the "all round" effect. Attach a hook to a slice of foam, hang it from a ribbon and create a flower sculpture to hang over a bed, in an alcove, or over a mantel. Brush a picture frame or the edge of a piece of mirror-glass with adhesive and press on flowers to make a floral border. This is a thrifty way of using small cuttings and snapped-off flower heads.

For the most delightful gift of all, take some ideas from the fresh flower section and shape a bouquet with dried flowers. All the rosebuds, camellias, and lilies-of-the-valley, dried in crystals or powder, are bouquets in waiting, their stems already wired.

A medley of dried flowers and seedpods, bunched in a deep basket, makes a pleasing, long-term feature indoors.

FLOWERS FOR AIR DRYING

Flower	Method	Notes
Acanthus	Hang in bunches	Tall arcs of pink, white and mauve flowers
Achillea or Yarrow	Upright, or in bunches	Small young flower heads are most successful
Agapanthus	Upright	Allow plenty of space between seedpods. Hollow succulent stems take weeks. Can be varnished
Alchemilla mollis	Upright or hanging in small bunches	Flowers gradually fade from lime green to mustard yellow
Allium species	Stand upright, supporting heads on racks, or flat in boxes	Wide range of sizes and colors including white, yellow and purple
Anaphalis	Hang upside down	Harvest before full-blown. Do not strip off foliage. Dries well
Astilbe	Hang in bunches	Long, feathery stems give a misty look to arrangements
Ballota	Dry flat	Cut when young and green. Keep leaves on
Bay leaves	Hang in bunches; can also be glycerined	The leaves turn brown after both drying and preserving in glycerin
Bells of Ireland (*Molucella laevis*)	Hang singly. Can also be preserved in glycerin	Strip off leaves. Fade to rich parchment color

Flower	Method	Notes
Bluebell (*Scilla*)	Seedpods – hang or stand upright in container to dry	Pick seedpods when becoming dry and papery
Broom	Hang dry stems	Can be twisted into a circle, tied firmly and hung on a hook to dry. When untied the stems will spring into gentle curves
Bulrush	Hang dry	Pick spikes when they are half developed, and spray with hair spray to prevent seeds from bursting
Burdock	Hang dry	The burs dry well and keep their color
Carrot	Hang dry leaves	Useful to give a feathery softness to small designs
Chamomile	Hang in bunches	Pick flowers before they reach full bloom
Chinese lantern (*Cape gooseberry*)	Hang dry seedpods in bunches	Cut when not all the calyces have ripened. Strip off leaves. When dried, hang some stems in sunlight to fade the bright orange "lanterns"
Chive	Dry flowers upright, with the stems through wire mesh to support the heads	Quick-drying in a linen closet helps to preserve color
Clarkia	Dry hanging or flat	Flowers dry extremely well and keep their color. Seedpods can also be dried
Clematis	Hang dry	The fluffy seedheads dry well. Spray them with hair spray to prevent them from disintegrating
Columbine (*Aquilegia*)	Hang dry	Wait to harvest seedpods until they open and start to curve outward
Cornflower	Hang in bunches in linen closet	Pink, blue and white flowers all keep color well
Cow parsnip (*Heracleum*)	Dry upright in containers	Harvest just as seedpods are formed
Dahlias (*Pom-pon*)	Stand in ¾-1 inch of water	Petals dry as water evaporates. Once flowers have dried, wash and thoroughly dry stems; otherwise dampness will spread
Delphinium and larkspur	Small young flower shoots can be dried in desiccant. Flowers (in a linen closet) and seedpods can be dried hanging or flat	Dry both flowers and seedpods. Pick flowers for drying when some buds are still unopened and dry in a heated room
Dock	Hang dry. Can also be glycerined	Seedpods change from lime green to deep red when dried

Fennel	Dry seedpods upright	Stick stems into dry foam or stand in roomy containers
Feverfew *(Matricaria)*	Hang in bunches	Pick just before fully open. The small white flowers are most effective arranged in clusters
Figwort *(Scrophularia)*	Hang upside down	Candelabralike clusters of seedpods are good for outlining
Foxglove *(Digitalis)*	Hang dry seedpods	Can snip off seedpods for use singly in small designs
Fritillary	Dry seedpods flat. Can then be varnished	Harvest early in the season
Globe artichoke *(Cynara scolymus)*	Dry individual stems by hanging, or standing upright in containers	The huge purple flower heads, like great thistles, dry well and turn glistening cream
Globe thistle *(Echinops ritro)*	Hang dry	Cut the steel-blue globular flower heads as they are just beginning to open. Later in their formation they disintegrate as they dry
Godetia	Dry seedpods flat	Harvest late summer, strip off leaves and dry clustered seedpods
Golden rod *(Solidago)*	Hang dry or upright in a container	Cut stems in various sizes. Sprays of the laterals can be snipped off for small-scale designs
Grape hyacinth *(Muscari)*	Dry seedpods flat or upright	Once dried, the seedpods look like papery bluebells
Grasses	Dry by hanging in bunches, upright in containers, or flat	"Top-heavy" grasses on slender stems are best dried flat
Gypsophila	Hang dry loosely in bunches	The minute flowers give a misty look to designs
Heathers	Stand in ¾–1 inch of water, or hang dry in bunches	Pick stems when flowers have just opened
Helipterum	Hang in bunches	Harvest the everlasting flowers just before fully open
Hollyhock *(Althaea)*	Seedpods – hang dry or can be dried flat	You can clip off and use individual seedpods in miniature arrangements
Honesty *(Lunaria)*	Seedpods – hang dry or dry upright	Cut stems when the center of the seedpod, like silver tissue paper, is still protected by two brown outer skins. Rub these layers between thumb and first finger when arranging the sprays
Hop *(Humulus)*	Hang upside down or can be glycerined	The pale olive flowers fade only slightly. These papery blossoms are useful in dried arrangements, providing neutral green coloring
Hyacinth – giant summer *(Galtonia)*	Hang in bunches	Harvest as last petals are falling – except in hot Indian summer, when they can dry on plant
Hydrangea	Stand flower heads in about 1 inch water. Separate florets can be powder dried	The petals dry as water evaporates. Once flowers have dried wash and thoroughly dry stems; otherwise dampness will spread
Jerusalem sage *(Phlomis)*	Dry flat	Seedpods dry quickly. Do not strip off the white woolly leaves
Knapweed *(Centaurea)*	Seedpods dry upright or hang in bunches	Gather as soon as seedpods open (midsummer – early fall). Cut some buds and varnish them
Larkspur *(see Delphinium)*		
Lavender *(Lavandula)*	Can be dried by hanging, or upright in containers, both dry or in a little water	Cut the stems before the flowers are fully open
Lovage	Dry seedpods upright	Stick stems into dry foam or stand in roomy containers
Love-in-a-mist *(Nigella)*	Hang dry	Harvest late summer. Remove most of the frondy leaves. The globular seed capsules look like purple-striped lanterns
Maize **(or Corn)**	Dry cobs in a linen closet, flat on racks. Spray with lacquer or paint with varnish once dried. Hang dry seedpods and leaves	The leaves dry to a pale avocado green color and look like crepe paper. The seedpods turn to pale parchment
Mallow	Hang dry	The long spikes of silver-gray seedpods are like closed-up stars
Marigold	Double flowers can be dried by hanging in a linen closet. Both double and single marigolds dry successfully in desiccants	Harvest flowers for drying just before they open fully
Montbretia *(Crocosmia)*	Dry long spikes of seedpods flat	The stems, with the small, pale cream seedpods, are good "points" in an arrangement
Mullein	Hang dry the long spikes of seedpods, or lay them flat. Large subjects like this are best dried in a linen closet	Such long spires are useful for large displays in public places

Narcissus	Cut the stems short and dry flat in a linen closet. Can also be powder dried	All snub-nosed, flat-flowered kinds retain their color well. Mount the dried flower heads on false wire stems
Peony heads (small)	Stand in about 1 inch water, or can be powder dried	Petals dry as water evaporates. Once flowers have dried, wash and thoroughly dry stems; otherwise any dampness will spread
Pampas grass (*Cortaderia*)	Hang dry	Should be cut before they start to shed. Can be kept in shape by spraying with ordinary hair spray
Pinks	Dry upright. Can also be powder dried	Cut before the flowers are fully open. To retain the natural color, it is important to keep them in a dark place
Poppy (*Papaver*)	Hang dry the seedpods	All kinds of poppies provide excellent seedpods that are invaluable in any collection of dried material. Colors range from pale cream, through pale green to deep purple
Rodanthe	Hang dry	These pinkish-white everlasting flowers retain their color well
Rosebuds (tightly closed)	Hang dry in linen closet. Can also be powder dried	Only small, very compact buds can be successfully air dried
Rue	Hang seedpods in bunches	Seedpods look like little wooden nuts
Sage	Hang seedpods in bunches or dry upright. Sprays of leaves and flowers can be dried in a linen closet	Leaves retain their aroma even after drying
Salvia	Dry flowers and seedpods upright or flat – flowers in an linen closet	Delicate spikes of seed capsules are attractive "neutrals" with more colorful subjects
Santolina	Dry upright in containers in a linen closet	If dried quickly the tiny flower heads will retain their bright yellow coloring
Sea holly (*Eryngium*)	Hang dry	Dry the pale blue round flower heads on the stems, with the leaves
Sea lavender or statice (*Limonium*)	Hang dry in small bunches	Harvest when flowers are just open. The everlasting flowers dry very well and the neutral shades give a misty, cloudy effect. Bright colors fade in strong sunlight – which can be an advantage

Senecio	Hang the flowers in bunches, and dry sprays of leaves upright in containers in about 1 inch of water	The silver-gray leaves retain the delicate tones well
Smoke tree (*Rhus*)	Hang dry	When dried, the seedpods look like a puff of smoke and soften the outline of an arrangement
Sorrel	Hang flower spikes in bunches. Can also be preserved in glycerin	Pick in full bloom preferably. If picked after formation of seed it will easily disintegrate unless sprayed with hair spray. Colors vary from pale green to dark brown
Spiraea	Dry upright	The white flowers of the variety Meadowsweet dry a pale parchment color
Stonecrop (*Sedum*)	Hang dry	The flowers dry to a subtle pinky-beige in soft sprays
Straw flowers (*Helichrysum bracteatum*)	Hang in bunches	These daisy-shaped everlasting flowers are among the most useful of dried plant material. If the brittle stems snap off, or where height is needed, attach false wire stems to the flower heads
Sweet William (*Dianthus*)	Dry upright, in dry containers or with a little water	Pick the flower heads just before they are fully open; they will dry into balls of muted colored flowers
Tansy	Hang dry the flower heads	The brilliant golden heads are especially useful for large arrangements
Teasel (*Dipsacus*)	Hang dry	The long stems with seedpods like hair brushes dry well. They are good subjects for drying
Thistle (Scotch or cotton)	Hang dry	Cut before fully open or they will explode
Sycamore	Hang dry	The "keys" should be cut before they are ripe
Willow	Hang dry stems after leaves have fallen	Can be twisted into a circle, tied firmly and hung on a hook to dry. When untied they will spring into gentle curves
Xeranthemum	Hang dry	Everlasting flowers with strong wiry stems which retain their color well after drying. Varnish some buds
Yarrow	Dry flowers or seedpods upright	Stick stems into dry foam or stand in roomy containers

DRYING FLOWERS IN DESICCANTS

Flower	Notes
Carnation	Cut before fully in bloom
Choisya (Mexican orange blossom)	Powder dry short sprays, flat in desiccant
Dahlias	Small pompon dahlias give best results. To dry very large ones, use sand as the desiccant
Daisy (lawn)	As the dried flowers are so small and brittle, it is best to mount them on a short length of stem wire before drying
Delphinium	Choose small young shoots. Large flowers and seedpods can be hang dried
Forsythia	Dry short sprays horizontally
Hollyhock (*Althaea*)	Powder dry open flowers or tightly closed buds. Double flowers dry more successfully
Hyacinth	Push a wire through stems before drying
Hydrangea florets	These separate florets can be very useful in miniature designs
Laburnum	Cut when not quite in full bloom. Lay short sprays flat. Retains shape and color well
Larkspur (see delphinium)	Dry small young shoots
Lilac	Dries very well. Cut before heads are fully open
Lily-of-the-valley	"Fades" to deep cream
London pride	Very useful shape for points in, for example, triangular arrangements
Marigold (*Calendula* and *Tagetes*)	Dry face down in the desiccant
Narcissus (Yellow-eyed)	Flowers can also be air dried flat
Orchid	The flowers retain their form, shape and color but lose the waxy sheen of the petals
Pansy	Mount flowers on short lengths of wire for stems before drying
Peony	Dry large, double flowers in sand. Can also be dried upright in a little water
Pinks	Single pinks should be powder dried. Double ones may be air dried, upright
Polyanthus (*Primula*)	Push a short length of wire into the calyx before drying
Rosebuds	Small, tight buds can also be hang dried in a linen closet
Snapdragon (*Antirrhinum*)	Pick off and dry each separate flower
Stock (double night-scented	Double flowers give best results
Sweet pea	Red colors strengthen considerably
Violet	Keeps color well
Wallflower	Sprays may be laid flat in desiccant
Zinnia	Dry face down. Small varieties dry most successfully

Flower	Notes
Broom	Cut when not quite in full bloom. Lay short sprays flat. Retains both color and shape well
Buttercup	Retains color well. Use lightweight powders such as borax to avoid crushing
Camellia	The double blooms dry best and retain their color and elegance. Leaves can be glycerined

A HARVEST OF SEEDS

There is no cause to sigh wistfully as flowers in gardens and fields around you fade; no need for a hint of self-reproach if you missed the perfect moment to preserve them. For as the petals fall, they give way to seedpods that in many cases are almost as attractive as the flowers themselves and a versatile second harvest.

Add to these a collection of delicate and ornamental grasses, a dramatic selection of vegetable and herb seedpods such as onion and lovage, for example – not forgetting the cereal harvest to be gleaned along the waysides – and you have a wealth of varied plant material that is rich in shape and subtle in color.

If you find the deep buttery golds, the sun-bleached creams, the misty blues and purples and the pure, unabashed silver (of honesty pods) a shade too understated, you can take matters into your own hands and dip-dye or spray paint seedpods to your own preference. I would not go so far as to say that these coloring techniques actually improve on nature, but they do give you a helping hand.

It takes a trained eye to see the decorative potential in a flower border or vegetable patch that has gone to seed. But once that eye is in training, even the trash in a friend's yard can be the most fruitful place on earth. Many is the time I have scrabbled among the prunings and thinnings of late summer and emerged triumphant with a handful of mallow stems, the seedpods like puffed-up stars; lupines, like pairs of silver-gray rabbits' ears; rue like clusters of tiny chestnut-colored wood carvings, and acanthus, like a sturdy ladder with rungs on each side of the central stem. It is amazing what some people will throw away.

PICKING YOUR TIME

As with any plant materials for drying, the harvest time is critical. It goes without saying that you should cut seedpods when they are outwardly dry, before, not after, a rainstorm. But that is not all. Seedpods are so fascinating because many of them significantly change shape, form and color as they develop, and it is up to you to keep an eye on them and choose which stage you wish to capture.

For instance, soon after the unremarkable mauve flowers of honesty fall, this plant develops clusters of flat, fleshy blue-green discs, each pair of which is supported on a short, spiky stem. At this stage the pods are very useful in foliage and "green" groups, bringing their unique shape and texture. As the seeds inside the discs ripen, the pods turn first purplish – when they blend beautifully with blackberry and red maple leaves – and then gradually bleach to beige. At this stage you can rub off the outer covers between your fingers – or leave them on the plant and nature will do it for you – to reveal the translucent silvery "moons."

If you grow honesty plants and can watch this evolution from your kitchen window, it is a good idea to cut some stems at each milestone. When the pods are still moist – at the green and purple stages – you can preserve them in glycerin solution. They emerge glistening bright and a deeper shade of the original color. Or, at any stage, simply hang them in bunches to dry.

SEED ART

Plants that form numerous flowers along a long stem are a treasure trove for the floral artist. Stately spires of delphinium, mullein, foxglove and delicate stems of parchment-colored grape hyacinth seedpods seem to give best results if they are dried flat; hollyhock, sage, dock, lupine, giant summer hyacinth and clarkia hanging upside down; and agapanthus, sometimes called lily-of-the-Nile, standing upright in containers. If you harvest agapanthus before the pods burst to release their seeds, be prepared for a shock. They explode with a crack that is almost indistinguishable from a rifle shot. Once dried, the pods can be painted with varnish.

The fact that some of these stems rise to towering heights does not mean they are suitable only for grand displays in churches and other public places. There is no need to use them as they are. Each stem can be snipped and clipped into a number of small elements. The short and slender side shoots are useful for "uprights," and individual seedpods, on their own stems or pushed onto false wire ones, can be delightful in small and miniature designs.

You can raid the vegetable patch as well as the flower garden for umbrella-shaped seedpods, and keep a look out for other species – giant hogweed, cow parsnip, fennel and yarrow – that grow freely in the wild. Angelica, caraway, carrot, chervil, lovage with its deeply ridged stems – all these umbellifer

seeds dry successfully when left standing upright. I keep a collection of used bottles and place one stem in each. They make an interesting group standing in an empty fireplace and have even been mistaken for a deliberate arrangement. Again, there are no constraints on any floral artist to use these huge protective heads as they appear on the plant. Snip off several of the tiny stems, cluster them together and push them into stem-holding foam as you build up a mini arrangement. Close up they look quite impressive.

PODS AND GRASSES

There is a whole range of plants that rewards the eagle-eyed arranger with a collection of various urn-shaped seed capsules that dry perfectly hanging in bunches. Gray-green poppy seedpods, probably the most "classic" shape of all, and prickly spiky teasels are among the best known. But there are many others, such as love-in-a-mist, the purple and green striped "lanterns" criss-crossed with a net of green whiskers; aquilegia with brown tulip-shaped pods; flag iris with long blue-gray tubes; campion, not unlike a smaller and less substantial version of poppy heads, and by far the most spectacular; and Chinese lanterns, which are bright orange on the plant and fade to a much more sociable shade of pale apricot. They all add interest and variety to dried arrangements or can be mixed with fresh flowers.

In an end-of-season frenzy of garden clearance, many people are apt to throw away a whole group of seedpods that have highly attractive chunky shapes. But once their petals have fallen, sunflowers, gaillardia, scabious and knapweed, vastly different in scale, leave a legacy of humpy centers that dry well hanging in bunches. Other favorites of mine are Jerusalem sage, with hoops of pale cream or green seed capsules around the stem, candytuft with whorls of seeds in clusters, and Greek spiny spurge, the seedpods closely resembling crisp brown lawn daisies. Many is the time I have braved unfriendly terrain in Greece to bring home a bunch of my annual souvenir.

The most graceful and slender seed carriers of all are wild and ornamental grasses, and cereals – wheat, oats, barley and rye – which sometimes you can gather from roadsides. Birds are helpful seed broadcasters and tend to scatter a crop far and wide enough to enable it to be harvested by flower craft enthusiasts without the need to trespass. All these materials are easygoing candidates for drying, and the option of either hanging them in bunches, standing them in containers or spreading them out on racks is dependent upon what is most convenient. Only those with the slenderest stems show any preference and these give the best results if dried flat.

A pitcher of dried grasses, wheat and wild oats, with a color burst of paper poppies, can evoke the golden days of harvest. A sun-bleached arrangement of seedpods emphasizing the fascinating variety of shapes can be a symphony of neutral creams and browns, the long-lasting equivalent of a foliage display.

DYEING AND SPRAYING

To the adventurous and imaginative floral artist, a collection of dried plant materials also presents a wealth of opportunity for experiments with color, as exciting a challenge as a piece of unbleached muslin to a batik enthusiast. Coloring plant materials, by dip-dyeing or spray-painting, can never be an exact art with precise recipes and methods timed to the minute. The color tones achieved will depend on the absorption properties of the various plants and whether they have retained any residual moisture.

For dip-dyeing I like to use natural dyes since they give more sympathetic results. Not having any woad (*Isatis tinctoria*) or dyers' poke-weed in the garden, I have turned instead to readily available fruit juices and flower petals. The juice strained from stewed blackberries, mulberries, raspberries, red currants and white currants (which makes a pale amber dye) work splendidly. The more you dilute the juice with water the softer the color will be. Onion skins simmered in water make a brown dye, and marigold petals (one cup of petals in one cup of water, simmered for one hour) give a pale yellow color.

I find that the heathery shades obtained with the black fruits and the soft pinks and corals made possible with the red fruits give the prettiest effects. You simply leave the juices to cool, pour them into jars and dip in the dried plants.

DYEING SEEDPODS

A collection of seedpods in all shapes and sizes dyed in a single color makes a charming group. It does not take much practice to realize that furry surfaces absorb dye most readily and emerge with the deepest shades. Teasels, Jerusalem sage, knapweed, grape hyacinth and many others have what I term medium absorption, and the shiniest and toughest examples – honesty "moons" and poppy pods – grudgingly accept the color. Honesty gives the most delightful effect, looking as if it has been spattered with colored rain.

Dip-dye the seedpods for a few seconds only, or else they will go soggy. Shake them gently over several layers of absorbent paper towels, and hang them to dry in a warm place, with protective newspapers underneath. When they are dry, store them in the usual way.

Spray-painting offers different possibilities – bright and shiny colors and a cloak of gold or silver for Christmas arrangements, for instance – and is yet another way of preserving flowers.

A GLEAM IN WINTER

There could scarcely be two more contrasting methods of preserving plant materials than drying and treating them in glycerin solution. Whereas drying – whether it is achieved in a free circulation of air, in powder or crystals, or by pressing between absorbent paper – draws out the natural moisture from the plant, the other method replaces that moisture with a heavy liquid that will not evaporate.

Not surprisingly, the results are quite different, too. As we have seen, dried flowers retain their color and shape but become crisp, papery and almost invariably have a matte texture. Materials preserved in glycerin, on the other hand, are soft, pliable, and positively glow. Leaves take on a sheen they may never have had in their fresh state; berries radiate color and gloss, and grasses and bracts have a velvety softness.

Where alternative preserving methods are recommended, it is interesting to treat some materials, such as Bells of Ireland and hydrangea, undeveloped honesty seedpods, dock and sorrel, grasses and cereals, bay leaves and others in each way, and note the different effects. Since there will be differences in both color and texture, it is useful to have examples of both methods in your collection. It is this contrast, matte with shiny, light color tints with deep shades, that gives the many forms of flowercrafts such lively visual interest.

Reputedly the art of preserving leaves was first practised by Alexandra, wife of Edward VII of Great Britain, whose fondness for gardens, and for flowers in the house, is well known. Like many who are new to the technique, the queen began by preserving sprays of beech. As her skill and enthusiasm grew so, it is said, did her ever more varied collection of leaves, which she liked to use as a background to winter flowers brought in from the hothouses.

Preserved leaves are naturally at their most useful in winter, when most gardens can offer little but evergreens and florists have very little foliage indeed. But there will be other times throughout the year when a spray of deep cream preserved eucalyptus, chestnut brown maple or yellowed birch can provide just the contrast needed to complement a design.

The process of preserving in glycerin gives

deciduous leaves many of the characteristics and colors they naturally acquire at the end of summer, in what a New England gardener friend calls the golden days. But you must not wait until the leaves are about to fall before you cut them. It will be too late when the sap is receding in the plant and, on the other hand, too early in spring, when the immature leaves lack the veins and fibers needed to support them.

Gather the branches of deciduous leaves in high summer (in practice, in most of the United States this means late June, July and most of August; in California it includes September, too). Evergreens

By contrast with plants that are dried with desiccants, leaves, berries and seedpods will shine when preserved with glycerin.

can be preserved over a longer period, but avoid the time when the tree or bush is forming new shoots of tender young leaves. At this stage they will be totally resistant to a sustaining drink of glycerin.

As you can use the glycerin solution over and over again – you simply need to strain it now and then, and eventually store it in a stoppered bottle – this method of preserving can be easily repeated.

If you have a garden, it seems almost wasteful not to keep your jars of glycerin constantly filled with leaves. And there is no need to call a halt when you have enough for your own use. To a flower arranger without access to laurel and bay, broom and bergenia, a bunch of preserved foliage is a welcome gift or an irresistible rummage sale purchase.

THE GLYCERIN TECHNIQUE

To make the preserving solution, mix one part of glycerin with two parts of very hot water and stir or whisk vigorously until it is thoroughly mixed. You might use a cup as the volume measure to start with, and then increase your supply as your enthusiasm and skill grow.

To preserve sprays of leaves, pour the solution into containers to a depth of 2 inches. As the stems take it up, you will need to keep the mixture topped up to this original level, so transparent containers such as preserving jars and bottles are an advantage. Large, fleshy leaves like bergenia, *Fatsia japonica* and fig are very slow to preserve on the stem, taking up to six weeks, and will sit in your glycerin solution for an undue length of time. To avoid this you can preserve separate leaves, two or three at a time, immersed in a flat dish of the solution. This cuts the time to 10-14 days. Only perfect specimens are worth picking, since flaws that may be scarcely noticeable in the fresh state will be magnified after preserving.

Cut sprays of leaves such as beech, silver birch or ash, that have graceful curves. Snip off damaged leaves and criss-crossing stems that confuse the main outline. Treat woody stems in the usual way, cutting them at a slant, making deep slits or lightly crushing the stem ends and scraping 2 inches off the bark. If you intend to preserve very thick, fleshy leaves on the branch, it helps to rub them first with absorbent cotton soaked in the glycerin solution. This prevents leaves at the extremities from drying out before the solution reaches them.

Stand the prepared stems in the solution, taking care that they actually reach the base of the container. Place individual leaves in a dish of the mixture and press them down to ensure complete immersion.

Patience is a useful virtue while the next stage takes its course, though the process never lacks interest. You can see the progress of the glycerin as it travels up the stems and reaches first the lower leaves and finally the topmost ones. When the leaves at the top, or the tips of large immersed leaves have turned brown and glossy, the time has come to remove them. Wipe the leaves and stems with a clean cloth to dry them and remove any oiliness. Tie stems into bunches and hang them upside down for a few days to "firm up" the tips. You can continue to store them this way or flat in boxes, whichever is more convenient.

Preserved plant material does not need to be arranged in water. Dry foam, a holder or crumpled wire mesh in a dry container is all that is needed to hold the stems. If you wish to use sprays of preserved leaves to complement fresh flowers in water or soaked foam, paint the stems with two coats of nail varnish to protect them from mold.

MATERIALS TO PRESERVE

The materials you preserve will naturally depend on what is available. If you have only one tree, shrub or flower bed in the garden, it is worth giving it a try. A few notes on preserving times and results will serve as a general guide. Sprays of beech leaves are preserved in four to seven days; ferns, oak, pittosporum, rose and hornbeam in 14 days. They all turn attractive and varying shades of brown. Variegated holly (about 21 days) turns chestnut brown; laurel and broom (14 days) almost black, and box (21-28 days) becomes beige.

Teasels harvested and preserved when they are green, pussy-willow, lime flowers (stripped of the leaves), sprays of blackberry and strawberry tree both in fruit, or spruce with cones take from 14-21 days, and add interest and variety to your collection.

You can capture the magic of berries and other fruits by preserving them on the stems in glycerin solution. Holly, rose hips, snowberries, skimmia, flowering quince, hypericum, euonymus, pyracantha – all give rewarding results. In general the color deepens and the gloss is retained. Give the shine an extra boost by spraying preserved berries with hair spray or painting them with varnish.

For short-term keeping, for Christmas and other winter decorations, maybe, simply paint fresh berries with varnish. This will extend their vase life by several weeks.

PRESERVING IN GLYCERIN

Material	Preserving days	Notes
Aspidistra	14-21 immersed, about 70 upright	Immersion gives better results
Bay (*Laurus nobilis*)	14	Leaves may turn dark olive green or chestnut brown
Beech (*Fagus*)	3-7	Cut in midsummer when fully mature
Bells of Ireland (*Moluceila*)	21	Preserve then hang upside down to dry
Berberis	21	Treat long stems. Leaves turn deep brown
Bergenia	14	Preserve the large leaves
Berries	14-21 depending on type	Treat sprays of berries and leaves together
Blackberry	21	Treat leaves and fruit
Box sprays	14-21	Small, compact sprays are useful fillers
Camellia (leaves)	14-21, immersed 21-28, upright	The flowers can also be powder dried
Choisya (leaves) (Mexican orange)	21-28	Treat only mature leaves
Clematis	14-21	Treat sprays of leaves and seedpods
Cypress	21-28	Treat flat fan-shaped sprays complete with "cones"
Dock	14	Cut small spikes of flowers
Escallonia	14	Sprays of minute leaves are ideal for shape outline
Fatsia (*Fatsia japonica*)	14-21, immersed about 70, upright	The evergreen foliage turns mid brown
Fig	7-14, immersed about 45, upright	Treat leaves singly
Gum tree (*Eucalyptus*)	14-21	Leaves turn gray-mauve
Hellebore (*Helleborus corsicus*)	21	Leaves turn light brown
Holly (*Ilex*)	21	Spray berries after preserving with hair spray
Hop (*Humulus*)	14-21	Sprays can be hang dried
Ivy (*Hedera*)	10-14, immersed about 21, upright, depending on length	Long sprays are best immersed. Berries can be treated also
Laurel (*Laurus nobilis*)	21-28	Store individual preserved leaves between blotting paper
Lime	18-21	Treat sprays of leaves and "keys"
Mahonia	21-35	Leaves darken progressively
Maidenhair fern (*Adiantum*)	14	Cut at any time except when new growth is forming
Maple (*Acer*)	14	Treat single leaves and clusters of "keys"
Mistletoe	14	Berries wither in time
Oak (*Quercus*)	14-21	Oak apples and acorns preserve well on the stems
Peony (*Paeonia*)	14	Leaves turn pale olive green with dark, contrasting veins
Pyracantha	21	Spray berries with hair spray
Raspberry	14	Cut young, perfect leaves
Rhododendron	21	Preserve sprays of leaves with young, tight buds
Rose briars	14	Young shoots of wild roses preserve well
Rose hips	14-21	Spray hips with hair spray after preserving
Rosemary (*Rosmarinus officinalis*)	14	Stems after preserving vary from dark green to silver gray. Leaves retain aroma
Mountain ash		Spray berries with hair spray
Shoo-fly plant (*Nicandia*)	14	Hang upside down to dry after preserving
Silver birch (*Betula*)	28	Cut with catkins on twigs. Use solution at boiling point
Snakeweed (*Polygonum*)	14-21	Fine sprays of dark red flowers keep color well
Snowberry (*Symphoricarpos*)	14-21	White snowberries deepen in color
Spanish moss	14	Gather before flowers open
Sweet chestnut (*Castanea sativa*)	10-14	Preserve in three stages with tiny chestnuts, and with and without catkins
Sycamore (*Platanus*)	7-10	Treat leaves and "keys"
Tree peony	10-14, immersed	Totally immerse short sprays or separate leaves
Viburnum	21	Leaves turn deep brown with olive brown undersides
White cedar (*Chamaecyparis*)	21	Leaves turn orange-brown. Do not remove from solution when they are just brown
Yew (*Taxus*)	21-28	Berries do not always preserve well

PRESSING FLOWERS

Anyone who frequents used-book stores has undoubtedly had the experience of opening a dusty volume and having it fall open to a page marked by a faded, brittle flower. Such occurrences add dimension to our lives, reinforcing our connections with the past and reminding us of the enduring allure of nature.

Drying plant materials by pressing presents an artistic challenge of two completely different kinds. First, the thrill of the hunt, for flowers and leaves that will retain their characteristic shape and, if not their original color, a hue that is pleasing. Then after the pressing process comes the creative joy of arranging petals, flowers, stalks and leaves into representational or abstract designs, for anything from a gift tag to a full-scale picture. One of the most evocative pressed-flower pictures I have seen depicted a thatched cottage surrounded by massed herbaceous borders, and a path overgrown with herbs, leading up to the entwined roses around the door.

PRESSING EQUIPMENT

The equipment you need for pressing is minimal, though you may choose to enlarge your kit as you progress. Highly absorbent paper is the main essential item. Blotting paper is hard to beat, though somewhat expensive now. Absorbent paper towels and tissue paper are suitable for small, thin subjects – be sure to choose a smooth paper without a waffle or ridged surface – and newspaper is serviceable for large leaves, which may be pressed under a mattress, sofa cushion or carpet.

Large books were traditionally used for pressing. Flowers may be enclosed directly between the pages, but check that the printing ink is not likely to transfer to the petals – not a pretty sight – and remember that the reverse can happen. It is a pity to spoil a valued book with chlorophyll stains. To be on the safe side, it is better to interleave the folios with absorbent paper.

As an advance on the book technique you can buy or make a flower press. This is a series of alternating sheets of flat cardboard and blotting paper enclosed top and bottom between boards and held together at each corner by thumbscrews. The trick is to exert similar pressure on each screw so that the materials are efficiently and evenly dried.

It is useful to have plastic tweezers to pick up subjects before and after pressing, and a small camel's-hair paint brush to move them about on the page. Handling delicate petals and flowers can all too easily cause bruising and result in an ugly brown stain.

The greatest asset to a beautiful collection of pressed materials must surely be a garden. There cannot be many craft activities that offer more sheer delight than wandering from plant to plant, cutting a spray of leaves here, a handful of flowers there, some buds, a few supple stalks. But a garden is not the be-all and end-all of the art. Take a calculating look at your house plants; there is plenty of potential there. A mixed bunch of flowers bought from a flower stand, or flowers in prime condition from a gift bouquet might also be used.

It goes without saying that plant materials must be dry and in perfect condition when you press them. If you gather flowers from a friend's garden or pick weeds in the countryside, give them a reviving drink when you get home. But keep the time between cutting and pressing as short as possible. Better still, take your books or presses with you on day trips and press on site. A word of caution: a gust of wind plays havoc with carefully arranged pages, so position yourself with this in mind.

CHOICE OF MATERIALS

Nature is at your disposal when it comes to deciding what to press, and it is fun to experiment with new subjects now and again. Flowers and leaves with thin tissue are likely to give consistently good results. Moist, fleshy leaves and flowers with thick centers (which prevent the absorbent paper from coming into contact with the petals) are a disappointment. Buttercup and clematis stalks are the best choice for natural curves in pressed flower designs, and it is worth trying to obtain these in preference to any other.

Snip off flat flowers such as borage, daisy and hydrangea florets just below the calyces. Pull off petals from thick flowers such as carnations and press them separately. One flower yields a multitude of

Kits for pressing flowers need only include a heavy book and absorbent paper, but an attractive flower press adds to the interest.

petals that can be reassembled more sparingly to create the illusion of a natural flower. Snip off leaves or leaflets for pressing separately.

Arrange flowers, petals and leaves of a single kind, or at least of equal thickness, on each sheet of paper. Take care that they are well apart, not touching and certainly not overlapping. A page spread with plant materials both before and after pressing is a beautiful feature of the craft.

Lower the covering sheet of absorbent paper or close the pages of a book carefully. The slightest whoosh of air disturbs the placement. Close the press, or weight books with a pile of additional volumes and leave the "pressing workshop" in a warm, dry room for at least six months, longer if possible. The longer you leave them, the less likely they are to fade when they are eventually exposed to light.

KEEPING THE COLOR

Color retention – in flowers more particularly than in leaves – can be a problem. I have found that changing the blotting paper layers two or three times during the first few days and transferring the subjects to dry paper helps considerably. And, of course, you can dry the papers for reuse.

Dampness is always an enemy. Once your materials are pressed and dry, layer them between fresh tissues and keep them in boxes in a warm, dry place. They are only too ready to reabsorb moisture, and that means discoloration and mold. A spoonful of silica gel crystals in a twist of muslin helps.

Blue has the reputation of being the most difficult

Flowers such as daffodils may be sliced in half for pressing, but the petals of bulky flowers, like carnations and roses, need to be separated from the heads. The calyces of daisies may need slitting to allow the heads to lie flat.

color to "hold," but there are many subjects that will retain this hue. Tiny sprays of speedwell pressed on the stem, single delphinium flowers, borage with their star-shaped calyces, alkanet (anchusa), lobelia, pansies and *Geranium ibericum* are a galaxy of sky-blue color.

Bright sunshine yellow flowers are a great success. Use buttercups pressed singly and as buds on the stem, marigold petals for later reassembly, celandine, Iceland poppies – complete flowers or separate petals – primroses, cowslips, chrysanthemum petals and daffodils. The technique with trumpet-shaped flowers in the narcissus family is to slit them through with a sharp knife and press each half, making two perfect shapes from one bloom.

Pink and red flowers tend to fade or deepen slightly. Bright red poppies soften to an attractive shade of old rose pink, and carmine old rose petals often deepen to strong purple. Fuchsia flowers make charming bell shapes, and wallflowers look like miniature flag iris. Saxifrage emerges like toy flowering shrubs.

White flowers usually take on varying tones of cream, with lawn daisies and Queen Anne's lace the exceptions. Used whole or snipped into florets, the latter – also known as wild carrot – looks just like delicate snowflakes. Green flowers and bracts such as the large trumpet-shaped tobacco plant and fluffy *Alchemilla mollis*, retain color well.

Many leaves deepen in color or turn varying shades of brown. Gray leaves such as senecio, anthemis, lavender, sage and cineraria are useful in design, and keep their neutral color well. But color is not everything in pressed flower work. Select leaves as much for the variety and interest of their shape. Clover, Queen Anne's lace, herb Robert, buttercup, achillea and aquilegia with their decorative cut shapes, slender wallflower and willow, feathery fennel and cosmos, young rose leaves and variegated ivy, fallen leaves of chestnut, beech, lime, in glorious shades of red and gold, all add natural variety.

DESIGNING ON THE FLAT

Think of a collection of dried and pressed flowers, leaves and stalks as a kaleidoscope of color and shape. Spread out several examples on a page, move them around with the tip of a small paint brush and a design will begin to emerge.

You might decide to criss-cross several stalks to simulate a basket and "fill" it with buttercups and daisies, some buds dipping over the rim as they would in a "3-D" arrangement. Or represent a circular table-top with poppy petals, adding a classic urn vase with three foxglove flowers and an exuberant bouquet of mixed flowers.

For a more abstract approach, select a graceful stem and, with no attempt at realism, arrange flowers and leaves on each side of it, a study in pleasing balance and shape.

You might like to create a scene – perhaps a waterside impression with striped grasses for rushes, slender willow leaves for sailing boats and Queen Anne's lace florets for the glistening foam on the water.

Whatever design you choose, rearrange the materials on a cardboard or colored paper background until you are satisfied with the effect. Work on one section at a time and dab each leaf, stalk or flower at intervals with paper glue on the point of a wooden toothpick. Position the material, cover it with a small piece of clean paper and press it gently in place. When you are overlapping petals and flowers – a daisy makes a pretty center for a composed flower – let one layer dry before imposing the next. Cover your completed work with a sheet of paper, then glass or a

board and some books to weight it down overnight.

Pressed flower designs always need protection from dust and scuffing. Cover greeting cards and tags with plastic wrap, pictures with glass or transparent iron-on mount. And to fight the battle against fading, display them away from direct sunlight.

SKELETONIZING LEAVES

Until you see a leaf that is skeletonized you cannot appreciate the intricacy of the vein structure. You can sometimes find these "outline leaves," stripped of the green plant tissue, under holly, magnolia or laurel trees and bushes, but perfect examples are rare.

They make a delicate contrast for blending with evergreens, a happy choice with dried flowers and a dainty "frill" surrounding a bouquet.

To strip evergreen leaves – the only ones with a sturdy enough structure – boil them for 30 minutes in 1 pint water to which you have added 4 ounces of enzyme detergent. Strain the leaves, rinse them under the cold tap and place them on newspapers. Use an old toothbrush to brush off the leaf tissue, working from the central vein and out to the sides. Rinse them again, dry them and press them between sheets of blotting paper for about 14 days. Store the leaves as for pressed flowers.

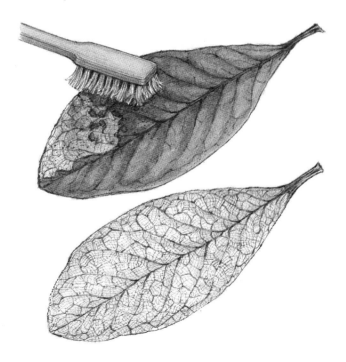

PRESERVING THE FRAGRANCE

Try to picture life on the large country estates and plantations of times past. Each of these homes generally had a special room for preparing preserves, liqueurs and beverages. Flowers and leaves hung in bunches to dry; others lay on wooden racks; shelves were filled with spices and fixatives, and there was a still to extract the essential oils from plant materials. In large households there might be a mistress of the potpourri, whose job it was to make the aromatic blends to perfume the rooms, some in priceless covered bowls and some sewn into fine linen bags.

Conditions have changed and there is no longer the need to strew our floors with dried mint, lemon balm, sweet cicely and other herbs to mask unwelcome odors, nor the necessity to place large bowls of pot-pourri at strategic points around a room, for the same purpose. But the delight of potpourri lingers still, and the gentle art involved in capturing the scents of a garden is more valuable to us than it ever was.

MAKING POTPOURRI

Potpourri is a medley of dried flowers, petals and herbs – some chosen for their fragrance and others for their shape and color. To those decorative ingredients are added grated dried fruit rinds, powdered spices, fixatives, essential oils and salt. With such a wide choice of components, you can vary basic recipes to suit your personality and preference, and create a blend that is uniquely your own.

There are two basic types of potpourri, the "moist" and the dry kinds. In the past, rose petals formed the foundation of both. Old-fashioned roses that grew in abundance in Elizabethan gardens were highly fragrant and rose petals have the advantage that they hold their fragrance for several months. American gardeners prized the Provence or cabbage rose, which they aptly named the hundred-leaved rose.

MOIST POTPOURRI

For the moist method the petals are partly dried until they are leathery, not crisp and brittle, and have been reduced to about half their original bulk. They are then packed into a jar with a lid, or a crock (never use a metal container) between layers of coarse salt, usually in the proportion of three cups of petals to one of salt. The mixture is left in a dry, dark, well-ventilated place for about 10 days. During this time the salt will draw out the residual moisture in the petals and the mixture will ferment, sometimes froth, and eventually form a block.

The fragrant block is broken up and other

ingredients added, to be sealed into a container and left to "work" for about six weeks. Opening the lid on the mixture to stir it daily, and noticing the gradual blending and mellowing of the fragrances, is one of the delights of the craft.

After this, other dried materials may be added, together with a few drops of essential oil. Another two weeks of harmonizing and stirring and the "moist" potpourri – now completely dry – made to a centuries-old method, is ready to use.

Above: *As soon as they are partly dried, rose petals can be "matured" in a container of salt.*
Right: *Spices and essential oils will enhance and fix the fragrant roses for potpourri.*

A traditional "dry" potpourri of pink carnations combines rose petals with cloves and cinnamon.

DRY POTPOURRI

The dry potpourri method is simpler. The rose petals and other plant materials are thoroughly dried until they are cornflake crisp, blended with the other aromatic ingredients, the spices, fixative and oils, placed in an airtight container and left to mellow for at least four weeks.

SOME FLORAL INGREDIENTS

A potpourri is essentially a personal pleasure. Choose the floral ingredients according to your own preference for fragrance and color. You may like to attempt a blend to resemble the perfume you wear, or color match a bedroom or living room scheme. And what more romantic a way could there be to preserve petals from a bridal or a Mother's Day bouquet?

To scent the potpourri mixture select from the petals of roses, carnations, flowering tobacco, honeysuckle, sweet pea, night-scented stock, orange blossom and clove carnation (the gillyflowers of old). To dry lily-of-the-valley, jasmine, sweet violet, verbena, hyacinth and lavender, first pick off the flowers and florets. And for the most charming touch of all, include some powder-dried rosebuds.

Many of these flowers are graced with both perfume and beauty, but there are countless other possibilities. Choose from forget-me-not, delphinium, larkspur, goldenrod and bachelor's buttons, and pansy, marigold, clematis and zinnia petals. They add color, texture and shape interest.

FRAGRANT LEAVES

Dried herbs are a traditional and delightful part of any potpourri, their soft green tones blending attractively with both vibrant and modest petal colors, their aroma linking and complementing the floral scents. There is much to choose from in the herb garden – rosemary, sweet marjoram, oregano, lemon verbena, scented geranium, lemon balm, mints of all types, lemon thyme, sweet basil, tarragon and sweet cicely.

In an ideal world the floral and herbal ingredients for potpourri would be dried on muslin racks on still, dry, warm days under the all-enveloping shade of a tree. To be more realistic, spread each flower and petal type separately on racks or trays in a warm, dry room, or in a cool oven. The more quickly the flowers dry, the more perfume they will retain. According to type and moisture content, flowers and petals will dry within two to 10 days in a warm room or linen closet. On trays in the oven at the lowest setting and with the door ajar they take only three to four hours. Stir them often and remove them from the oven as soon as they feel crisp.

You can dry sprays of herbs hanging in bunches – in the linen closet again – or strip the leaves from the stems and dry them flat.

A thoroughly modern way is to dry herbs in a microwave oven. Spread herb sprays or leaves on three layers of absorbent paper, cover them with some more paper and dry them in ½ ounce batches in the oven at 100 percent (high) for three to four

minutes. Allow them to cool completely before blending them with other materials.

When fresh herbs are unobtainable, you can use commercially dried ones, but make sure that they are "fresh" enough to have retained their color and aroma. Long-term keeping in a warm and steamy kitchen robs them of their volatile oils and effectiveness.

THE ADDITIVES

Raid the spice rack for powdered mace and ginger, and grind or crush whole allspice, cardamom, coriander, fennel and cumin seeds and cinnamon sticks. For all but the daintiest of sachets (when the sharp edges would be an annoyance) use whole cloves and star anise.

Dried citrus rind adds a sharp aroma that contrasts well with sweet floral scents. Thinly pare the rind of oranges, lemons and limes and dry the strips in a cool oven. Use thin strips or grind them to a powder in a blender or food processor.

The fixatives are a vital element in the making of potpourri. These are the ingredients that absorb and hold the scent of plants. There are some evocative names among traditional fixatives – frankincense, myrrh, sandalwood and patchouli. Those most used now are powdered orris root and gum benzoin, which you can buy at herbalists and drug stores.

Essential oils are extracted from strongly perfumed flowers and herbs and may be considered as potpourri seasonings. You can use them "like with like" to strengthen the existing fragrance – a few drops of oil of roses to enhance rose petals – or to add fragrance to blends chosen more for texture and color than for fragrance. And when eventually the magic aroma of a potpourri fades and becomes musty, you can instantly revive the mixture with a few drops of pure flower oil, stirred in.

Good herbalists sell a delightful range of essential oils. Among the ones I would choose are oil of rose, rose geranium, rosemary, citronella, lemongrass, pine, tangerine, vetiver, musk, violet, lavender and lemon verbena.

The potpourri recipes that follow are selected as a springboard to many others you will want to make from the ingredients available to you. As a general guide, to four cups of fresh flower petals and herb leaves, allow just over one cup of salt (if you are making potpourri by the moist method) at least 2 tablespoons of spices, one-third cup of fixative, and two to three drops of essential oil.

HOLDING THE FRAGRANCE

Gathering an aromatic harvest of flowers and leaves to dry is the first stage in a craft that branches off in many directions. If you have a workbox full of bits and pieces of material, ribbons and lace, you have all it takes to make sachets, bags and pillows filled with fragrance. If even simple sewing is not one of your interests, you can keep alive a delightful tradition and tie lavender stems into crisp bundles. And if the idea of fragrant fruit appeals to you – and clove apples and clove oranges are surely irresistible – wrap these spicy pomanders in potpourri. It's a notion that captures the essence of flowercrafts.

POMANDERS

In the 15th and 16th centuries pomanders – known as clove oranges – were carried to cloud unpleasant odors, ward off fevers and generally sweeten the air. It would have been the height of luxury to consider that they were also delightfully ornamental. Now that their original reason for existence has evaporated in a wave of health and hygiene awareness, we can enjoy them for the pure pleasure of their aroma – reminiscent of a waft of orange blossom on the warm summer air – and for their decorative properties. Studded with nut-brown cloves, sometimes with dried flowers and grasses too, and prettily ribboned, pomanders look good piled in a bowl, hanging in a bedroom and, at Christmas time, on the festival tree.

For the spherical base you can use any long-lasting fruit. Thin-skinned oranges are traditional, but green apples such as pippins and Granny Smiths were our choice. Small grapefruit and limes are other possibilities.

An unblemished green apple that is not too ripe makes a good base for a lightly perfumed pomander.

THE "CLOVE APPLE"

Any green apple may be used for the pomander base. The cloves can be packed closely on the fruit, even though apples do shrink more than oranges.

1 large green apple
12 inches of ½-inch-wide tape
a few needlework pins
1 wooden toothpick
2 tbsp. whole cloves
1 tsp. dried orris root powder
1 tsp. powdered cinnamon
1 small metal curtain hook or loop
24 inches of ½-inch-wide satin ribbon

1. Mark the apple into four equal sections, or quarters. Cut the tape in half. Place one piece around the apple from the stalk end, underneath and back to the top and pin it in place to divide the apple – visually, that is – in half. Pin the second piece of tape crossing the first, so that you have four quarters.

2. Use the toothpick if necessary to pierce holes in the skin. Working in one section at a time, press the cloves into the apple in a line closely following the line of the tape on the left of the section and then on the right. Keep the rows even and gradually fill in the section toward the center, until it is completed. Fill in the other sections in the same way.

3. Put the orris root and cinnamon into a paper bag and shake the studded apple in the mixture. Close the bag tightly, pressing it around the apple, and put it in a box to mature for two weeks.

4. Unwrap the pomander, remove the tape – which will look very grubby – and press the hook or loop into the top. Pin ribbon in place of the tape and tie a bow as a final flourish on the top.

Spicy variety

Variety being the spice of life, it is fun to take the original concept, of packing a piece of fruit with spices and leaving it to dry, and to develop it in different ways over a period of time. Instead of dividing the fruit into sections with ribbons, section it with rows of dried flowers or grasses. When you take off the masking tape, fill in the lines with dried santolina flowers or helichrysum buds. Or use these and other dried materials from your collection to make other patterns. Concentric circles or wavy lines, for instance, will make your pomander distinctive, and perhaps a conversation piece.

It is not quite so easy to hold the fruit in your hand and stud it with brittle flowers, so pierce the fruit with a knitting needle, stick it into a piece of modeling clay on a saucer and you can rotate the design and have both hands free.

Closer to the original concept, play down the design element and maximize on the aromatic spices. Stud oranges or apples all over with cloves (you will need more for each, of course) and roll them in spices. Add grated nutmeg, ground allspice and powdered cloves to the mixture for extra piquancy. Wrap the fruits, leave them to mellow, and then pile them pyramid fashion in a bowl. Alternatively place two or three – more sparingly – on a large bowl of lavender flowers or potpourri to make an attractive feature that will last.

Pomanders and potpourri were made for each other. Here is another spicing trick. Roll the pomander first in spices, then in potpourri before setting it aside. The petals will cling delicately to the cloves, and look attractive when the pomander is placed in a bowl.

LAVENDER BOTTLES

For centuries people have delighted in making bundles, bags and sachets of sweet-smelling flowers to scent their linen closets. There is a strongly sensual pleasure in taking out clothes and household items fragrant with the heady scents of summer: a practical application too. Moths do not find the powerful perfume at all attractive, and so will not be tempted to lay eggs among clothes.

Lavender bundles – or lavender bottles, to use the traditional and descriptive term – are simple to make and compact enough to tuck into a drawer, suspend in a clothes closet or hide in a trousseau suitcase. They also make delightful gifts and profitable rummage-sale items.

Cut fresh lavender flowers on a dry day, cutting the stalks as long as possible. Gather the stalks into bunches of about 20-24, depending on thickness, with the flower heads level. Tie the stalks together just under the heads with fine thread and bind the two ends around and around the flowers to enclose them in a tight web. Secure and tie the thread ends and cut them off.

Bend the stalks just below the flower heads, upward and over the binding and the flowers so that the lavender is enclosed completely. Neatly bind the stalks together, well above the tips of the flower heads – the design will now strongly resemble a baby's rattle. Cut off the stalk ends to make a neat finish.

Cover the binding with a generous bow of narrow pale mauve, white or mint green satin ribbon. Or – as in our photograph – weave ribbon under and over each group of four stalks covering the flowers. The ribbon binding not only adds a look of luxury. It ensures that once they are dried and have consequently shrunk and fallen from the heads, the flowers cannot fall out between the stalks.

PERFUMED PILLOWS

Heart-shaped sachets, drawstring bags trimmed with flowers, patchwork pockets, lacy pillows, tartan triangles, come in all shapes and sizes, and they can all possess a soothing aroma.

2 tbsp. dried lavender flowers
2 tbsp. dried rosemary leaves stripped from the stalks
10 cloves, lightly crushed
½ tsp. powdered dry orange

Mix the ingredients together and spoon them into a bag or sachet. A small pillowcase made of muslin placed inside a linen case will retain the fragrance.

Variations
Any medley of fragrant flowers, petals and leaves, whether it is a full-scale potpourri or an aromatic duet, can be enclosed in the fabric – in small sachets to slip into a drawer, bags to hang with a suit or dress, or pillows to send you to sleep on a waft of perfume.

Throughout the ages some herbs have been especially associated with inducing sleep – so when making a mixture for head-rest pillows, try to include dried leaves of angelica, sage, dill weed, bergamot, lemon balm, tarragon, scented geranium or rosemary, and chamomile, lavender or elder flowers.

ROSE POTPOURRI

This fragrant mixture is made by the moist method.

10 cups fragrant rose petals
½ cup block salt
½ cup table salt
2 tbsp. ground mace
2 tbsp. ground cloves
2 tbsp. ground allspice
2 tbsp. ground nutmeg
1 stick cinnamon, crushed
¾ cup dried, powdered orris root
1 cup dried red rosebuds
4 or 5 drops oil of rose

Dry the rose petals until they are leathery. Mix the block salt and table salt. In a glass or earthenware container, make layers of petals and salt, beginning and finishing with petals. Cover and leave for 10 days in a dry, airy room, stirring the mixture each day. The petals will ferment and by the end of this stage of the process will form a crumbly block.

Break up the petals and stir in the spices, cover and set aside for six weeks, stirring frequently. Stir in the oil of roses. Transfer the now dry potpourri to bowls, scattering a few dried rosebuds on top.

LEMON-SCENTED GOLDEN POTPOURRI

The sharp citrus aroma and the brilliant yellow coloring of this potpourri makes it especially refreshing.

1 cup dried lemon verbena leaves
1 cup dried lemon balm leaves
1 cup dried forsythia flowers
1 cup dried marigold petals
thinly pared rind of 1 lemon, cut into fine strips
⅓ cup dried, powdered orris root
1 cup dried chamomile flowers
6 drops oil of lemon verbena

Mix together all the ingredients except the oil of lemon verbena. Put them into a container with a lid, and set aside in a warm, dry, airy place for six weeks. Stir or shake the potpourri daily. Stir in the oil.

KEEPING THE FRAGRANCE

Traditionally, open bowls are used for the display of potpourri. Not only can you readily appreciate the colors and texture of the dried plants, you can also

Made by the "dry" method, lemon-scented potpourri gives lasting fragrance with a difference.

stir them with your fingers as you pass, and release perfumes as strong as those of a herb garden in the evening. Porcelain bowls, pedestal cake stands, bon-bon baskets are charming containers, but somewhat wasteful. The fragrance will more readily evaporate with constant exposure to the air. Check it frequently – be sure to remember how lingering it once was – and stir in a few drops of oil from time to time.

Pretty glass jars, heart-shaped bowls with lids, and ginger jars are thrifty alternatives. Take the lid off the container just before friends are coming, to offer them a fragrant welcome.

Potpourri has a natural affinity with aromatic pomanders (there are ideas for pomanders on the preceding pages) and with dried flowers. Fill a glass container – a cylindrical vase or a deep rectangular bowl – with potpourri and use this as stem-holding material. Arrange dried flower stems in the mixture and you have a perfect visual partnership. If the container has a wide neck, bury a piece of foam near the top to give the stems a better grip.

Potpourri can be used for wall, table and hanging decorations. Buy dry foam rings, heart shapes, strips and spheres and brush them liberally with glue, then press on the dried leaves and petals. Tie the potpourri ball with ribbons to hang like a pomander.

You could fill a deep, round basket with potpourri, or cheat and cram the basket with crumpled tissue paper, with a thin layer of flower petals on top. Make a dried flower ring to rim the basket and you have a delightful small arrangement for a bedroom or hearth.

In a beautiful house near London, in England, I saw bedroom hearths used to display potpourri. Each small blackened iron grate was heaped with petals gathered from the rose garden beneath the windows. It looked and smelled like the luxury it was.

97

THE LANGUAGE OF FLOWERS

For hundreds of years a single flower or a simple bouquet could speak volumes between two people. The language of flowers has its roots deep in the past. In the 19th century many people conveyed their messages with flowers. When a single flower was given and inclined to the right, the pronoun "I" was indicated, but "thou" when it was tilted to the left. If the flower "asked" a question and the recipient touched it to her lips, the answer was yes. If she pulled off a petal and let it flutter to the ground, the answer was no.

Our illustration shows the breadth of feelings that can be conveyed in a nosegay. You can use our own extracts from the dictionary to compose bouquets to offer for an anniversary, as a thank-you gift, an expression of sympathy, or indeed love and affection. Just one word of warning: it is a good idea to copy out the specific flower meanings and enclose them on a card. There may be several, sometimes conflicting, meanings attached to some flowers.

An eloquent bouquet uses ranunculus to say "I am dazzled by your charms," ox-eye daisies for "here is a token of love," and bluebells "to show constancy." The message ends with "do not refuse me," shown by the presence of California poppy.

LOVE, UNDYING LOVE

Amaranth, globe	*Undying love*
Ambrosia	*Love returned*
American linden	*Matrimony*
California poppy	*Do not refuse me*
Canterbury bell, blue	*Constancy*
Carnation, pink	*Woman's love*
Chinese primrose	*Lasting love*
Chrysanthemum, red	*I love*
Forget-me-not	*True love*
Gorse	*Love for all seasons*
Lemon blossom	*Fidelity in love*
Lime	*Conjugal love*
Myrtle	*Love*
Pink, red double	*Pure and ardent love*
single	*Pure love*
Rose	*Love*
bridal	*Happy love*
Rosebud, red	*Young love*
Silene, red	*Youthful Love*
Tulip, red	*Declaration of love*
Violet, blue	*Faithfulness*

AFFECTION, FRIENDSHIP

Acacia	*Platonic love*
Bluebell	*Constancy*
Daisy wreath	*Childhood affection*
Geranium, oak-leaved	*True friendship*

Heliotrope	Faithfulness
Hyacinth, blue	Constancy
Immortelles	Never-ceasing remembrance
Ivy	Friendship
Pear blossom	Affection
Periwinkle, blue	Early friendship
Valerian, pink	Unity
Veronica	Fidelity
Wallflower	Fidelity in misfortune

SORROW, GRIEF

Aloe	Affliction, grief
Anemone, garden	Forsaken
Balm	Sympathy
Columbine	Anxious and trembling
Harebell	Grief
Hyacinth, purple	Sorrow
Marigold	Grief
Poppy, red	Consolation
Tulip, yellow	Hopeless love
Yarrow	Solace
Yew	Sadness

GRATITUDE

Agrimony	Thankfulness
Bellflower, white	Gratitude
Campanula	Gratitude

UNDERSTANDING AND HOPE

Almond	Hope
Elderflower	Compassion
Geranium, scarlet	Comforting
Rose, stripped of thorns	There is nothing to fear
Snowdrop	Consolation and hope

REJECTION, HATRED

Basil	Hatred
Birdsfoot trefoil	Revenge
Bur	You weary me
Burdock	Touch me not

Carnation, striped	Refusal
white	Disdain
Chrysanthemum, yellow	Slighted love
Garden anemone	Forsaken
Larkspur, pink	Fickleness
Lobelia	Malevolence
Love-lies-bleeding	Hopelessness
Pink, variegated	Refusal
Indian double	Aversion
Rue	Disdain
Snapdragon	No
Tansy	I am your enemy

ADMIRATION

Alyssum, sweet	Worth beyond beauty
Fennel	Worthy of praise
Garden ranunculus	You are rich in attractions
Garden sage	Esteem
Geranium, apple	Present preference
Orange blossom	Your purity equals your loveliness
Pansy	You occupy my thoughts
Peach	Your qualities, like your charms, are unequaled
Pineapple	You are perfect
Ranunculus	I am dazzled by your charms
Rose, Austrian	You are all that is lovely
full red	I admire your beauty
centifolia	Your charms are manifold
Salvia, blue	You are wise
Sweet William	Gallantry, a smile

OLD-FASHIONED VIRTUES

Acacia, pink	Elegance
American cowslip	Divine beauty
Ash, mountain	Prudence
Azalea	Temperance
Bachelor's button	Celibacy
Bel orchis	Industry

Centaury	Delicacy
Broom	Neatness, humility
Clove gillyflower	Dignity
Cowslip	Divine beauty
Dock	Patience
Fuchsia, scarlet	Taste and gracefulness
Hibiscus	Delicate beauty
Lilac, white	Purity, modesty
Marshmallow	Mildness
Mint	Virtue
Mugwort	Tranquillity
Mullein	Good nature
Pasque flower	Without pretension
Rose, campion	Gentility
Roses, crown made of	Reward of virtues
Rose, single	Unaffected simplicity
moss	Superior merit
Sage	Domestic virtues
Salvia, red	Energy
Speedwell	Female fidelity
Star of Bethlehem	Purity
Violet, white	Innocence, modesty
Water lily	Purity of heart

OLD-FASHIONED VICES

Abatine	Fickleness
Bramble	Envy
Bugloss	Falsehood
Cherry blossom	Insincerity
Deadly nightshade	Falsehood
Fool's parsley	Silliness
Foxglove	Insincerity
French marigold	Jealousy
Mesembryan-themum	Idleness
Narcissus	Self-interest
Primrose, evening	Inconstancy
Rose, yellow	Infidelity
Sunflower	Haughtiness

Chapter 3

ARTIFICIAL FLOWERS

It is the most natural thing in the world for any artist to want to capture the beauty of garden and countryside flowers, and convey something of the emotions and moods they arouse. Flowers of every kind, the primary-colored flowers of spring, delicate, dew-splashed rosebuds, bright golden sunflowers, or bold, brilliant red poppies splattering color through a crop of corn have inspired artists and craft workers throughout the ages. Oils, watercolors and acrylic paints; clay, wood and stone; wax, wire, plastics and paper; silk, nylon and cotton materials; almond paste, yeast dough and "play-dough" – there is scarcely a material that has not been used to portray love and appreciation for the flowers around us.

In this section we have gathered together a harvest of designs for flowers you can make quite simply at home from paper and different types of fabric. And we hope the accompanying photographs will stimulate a wealth of ideas for the many ways you can display and use your crop of handcrafted blooms.

Paper is the most versatile of materials for making flowers; high-quality crepe paper – the kind that can stretch well – being the most useful of all. Our designers prove just how many shapes and effects can be achieved with it, from the smooth, sculptured look of tulips, and the exaggerated, generous size and floppy habit of huge field poppies to the trumpet-profile of crisp golden yellow and snow-white narcissus and the deep bell shapes of fuchsia. We include some fine, spider chrysanthemums and delicate roses, showing how such flowers can be made to look realistic and attractive.

In these practice sessions it is fun to experiment with coloring techniques too. You can smudge pale colored paper petals with crayon, dabbed and smeared on with absorbent cotton. The tulips shown on page 101 demonstrate just how effective this color application can be, adding not only a hint of shaded color but an illusion of shape and extra fullness. By contrast, with the utmost care and wearing protective gloves, use a weak solution of household bleach to subtract color from petal papers in deep, dark shades. Pale bleached streaks give a dramatic two-tone or bicolor petal effect – the cream and blue anemones on page 113 demonstrate this well. Try also dipping petal and stamen tips in bleach for color-lightened edges or spattering flowers with diluted bleach for specks and flecks.

FABRICS OF ALL KINDS

You have the chance to introduce a variety of texture with a range of materials – the delicate, pale-colored art-deco-inspired herbaceous flowers (page 124) were made of nylon, and the shamelessly shiny sunflowers were made of plastic (page 121). What many fabrics have in texture and sheen, however, they lack in malleability; very few can be twisted and stretched like crepe paper. This means that you need to apply hidden supports in the form of wires around the edge or through the center; only then can you coax the petals into the shape of, for example, camellias, iris or poinsettias.

Paper and fabric flowers have one common bond – they are all usually mounted on wire stems, or thin slivers of wood if you do not wish them to be bendable. Mounting the flower head firmly onto the false stem wire, wrapping the petals around the stem and then binding it along its length with florist's self-adhesive binding tape are all stages in flower making that take time and attention to perfect. Time devoted to practicing these stages is well spent in terms of finished appearance.

Paper tulips will bring springtime freshness to your home (see overleaf).

TULIP TIME

Tulips will add a perennial touch of spring to your home. Our design was inspired by the tulip variety Angelique, the white paper petals softly tinted green and pink with crayon. Deeper shades, simulating the stripes, splashes and flecks that give these stately spring flowers such appealing variety, may be added with poster paint.

Materials you will need
(for 12 flowers)
1 package heavy craft flower paper, white (or chosen petal color)
1 package heavy craft paper, leaf green
1 package crepe paper, leaf green
1 package crepe paper, black
roll of fine wire, for binding
florist's stem wires
colored crayons, green and pink
absorbent cotton balls
paste

Cutting and using the templates
Trace the patterns of the leaf and petal onto cardboard and cut out the templates. Place the petal template straight on the grain of the white paper, draw around it and cut six petals for each flower you wish to make. You will find that to save time you can cut through four layers of the craft paper, and six layers of the crepe paper. From the green craft paper cut two leaves for each flower.

Making the flower centers and stamens
Make the flower centers from green crepe paper. For each one, cut a strip of paper 3½ inches long and 2½ inches wide. Fold it in half across the grain. Fold over the top 1 inch of a stem wire to make a tight loop. With the folded edge of the strip facing upward, roll it tightly around the wire loop so that it is completely hidden. Secure the end with a dab of paste. Shade the top with a light green crayon. To achieve a natural effect, rub the crayon onto a ball of absorbent cotton and then dab this onto the paper.

To make the stamens, cut a strip of black crepe paper through all thicknesses of the pack about 2 inches wide. Make straight cuts along one edge 1½ inches deep and ³⁄₁₆ inches apart. Round off the top corners. Then cut the strip into lengths of six stamens for each center. Twist each stamen separately,

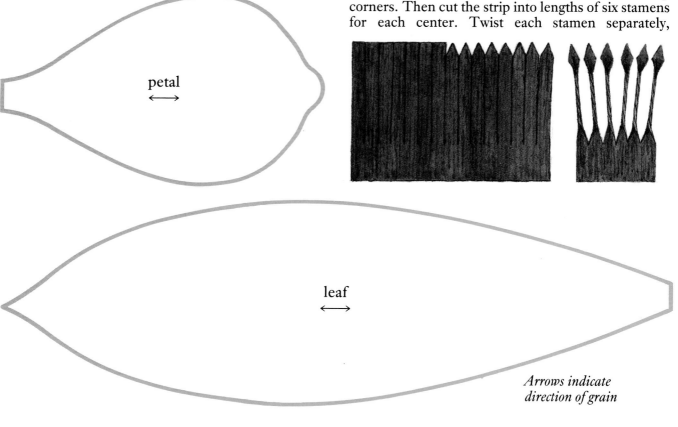

petal
←→

leaf
←→

*Arrows indicate
direction of grain*

starting ¼ inch from top. Wrap a stamen strip around a green flower center. Space out the stamens evenly with the black tips ½ inch above the green roll and paste them in place.

that bulge to lie just above the fastening. Place the remaining three petals to form a second row. Make sure that the centers of this layer cover the overlapping edges of the first layer of petals, and that they are all level at the top. Bind the petals tightly around the base. If you wish to strengthen the stem or make it thicker, insert another stem wire. Push it into the base of the flower and twist the two wires around each other in two or three places.

Treating the petals

If you wish to color the petals, do this before shaping and fitting them. Dab the base of each one very lightly with the absorbent cotton colored with the green crayon. Apply the color gradually, building it up until you achieve the depth of color you wish. Streak the top of the petals lightly with absorbent cotton rubbed on the pink crayon.

To shape each petal, hold it about 1½ inches above the base between the forefingers and thumbs of both hands. Gradually stretch it carefully and evenly all around the width to make a full cupped shape.

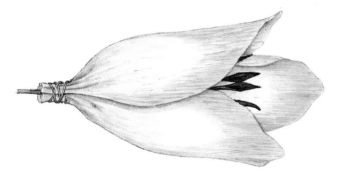

Assembling the flower

To assemble each flower, arrange three petals, evenly spaced, around the green and black center. Bind them on tightly with binding wire, allowing the parts

Covering the stem

From the green crepe paper cut a strip measuring ¾ inch wide through all thicknesses of the pack. Secure the end of the strip with a dab of glue just above the binding wire securing the base of the petals. Bind the paper strip tightly over the base of the petals and down the stem, overlapping it neatly and keeping it taut.

Inserting the leaves

Toward the base of the stem, insert the two leaves, on opposite sides, binding them in place with fine wire. Continue binding the stem wire with the green paper strip. At the end of the wire, cut the paper strip and secure it with a dab of glue. Bind a second strip to cover the first in the same way.

Final touches

Insert your fingers and thumb into the flower and stretch the petals a little more to give the flowers the characteristic cup shape. To store the flowers before arranging them, pack them in a box between layers of tissue paper, or push the stems into a block of dry foam.

A piece of foam in the base of a deep vase will hold the stems in place when you arrange them. Wire tulip stems have a great advantage over natural ones – you can ease them into gentle curves or have them completely vertical – just as you wish.

MEXICAN POPPIES

Flowers as big, bold and beautiful as these paper poppies are a passport to festivity. Make them as an eye-catching feature when you are planning a special party; stand them in a stone pot outdoors on a balcony or patio, to be the most exotic blooms in any garden; or cluster them in a room corner to brighten dull winter days. Wherever they are, these larger-than-life flowers will steal the limelight.

Materials you will need
(for 6 flowers)
cardboard, for petal template
1 package crepe paper, bright orange
1 package crepe paper, black
1-inch-diameter buttons
6 florist's stem wires
11 feet thick wire, for stems
roll of fine wire, for binding
1 knitting needle
florist's green binding tape, to bind stems
absorbent cotton balls

Cutting and using the templates
Trace the pattern of the petal onto cardboard and cut out the template. For each flower, cut eight pieces of orange crepe paper, 5½ x 4½ inches, and, using the template, cut out the petal shapes.

Making the flower centers and stamens
Push a stem wire through the holes in a button and twist the ends underneath. Cover the top of the button with a small piece of absorbent cotton. Cut a small square of black paper, cover the top of the button, shaping the cotton to make a mound, and bind the paper beneath with fine silver wire.

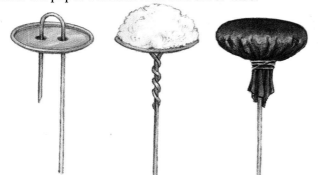

Cut a strip of black paper 13 x 4 inches and fold it in half along the length. Cut narrow slits 1½ inches long from the fold through both thicknesses all along the length of the strip. Gather the strip around the button and wire it beneath the button to the stem.

Hold the knitting needle against the outside of the cut strip and wrap the slits in the paper around it, several at a time, to give a ruffled effect.

Assembling the flower
Bind on the first four petals one at a time, spacing them equally around the stem. Bind on the next four petals at equal intervals between them, then bind them all securely onto the stem. Using your forefingers and thumbs, stretch the lower part of each petal to give the flower a natural cupped shape.

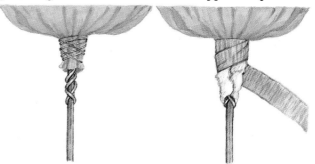

Ruffle the top edge of each petal: hold the edge between the forefingers and thumbs of each hand. Push the paper away from you with one hand and pull it toward you with the other, stretching it carefully. Work all around the top edge of each petal.

Making the stem and calyx
Cut 20 inches of the thick wire and push it up through the wired-on stamens to the base of the button. Twist the two free ends of the stem wire tightly around the stem wire. Wrap a little absorbent cotton around the stem beneath the button to simulate a calyx. Cover the cotton with binding tape and then carefully bind the full length of the stem.

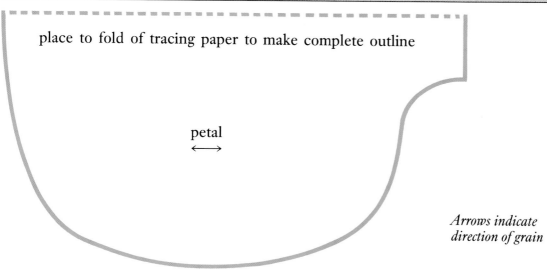

place to fold of tracing paper to make complete outline

petal
$\longleftrightarrow$

*Arrows indicate
direction of grain*

GOLDEN DAFFODILS

Our sunshine-bright narcissi are made from one of the simplest of paper flower patterns, and you do not need particularly nimble fingers to assemble them. They could be displayed in a cylindrical white vase, a piece of bleached driftwood or a shallow basket.

Materials you will need
(for 12 flowers)
cardboard, for flower and leaf templates
1 package heavy craft paper, yellow or white, for
    petals
1 package heavy craft paper, lemon, for trumpets
small amount crepe paper, beige, for sheath
small amount crepe paper, yellow
1 package crepe paper, leaf green
roll of fine wire, for binding
florist's stem wires
colored crayon, green
absorbent cotton balls
paste

Cutting and using the templates
Trace the patterns onto cardboard and cut out the templates.

For each flower cut six petals, one trumpet, one sheath and one or two leaves.

Making the flower centers and stamens
To make the center, bend over the top of a stem wire for 1 inch to form a narrow loop. Cut a ¼-inch strip of yellow crepe paper. Secure one end to the top of the loop and bind it carefully around to about 1 inch below the loop to conceal it.

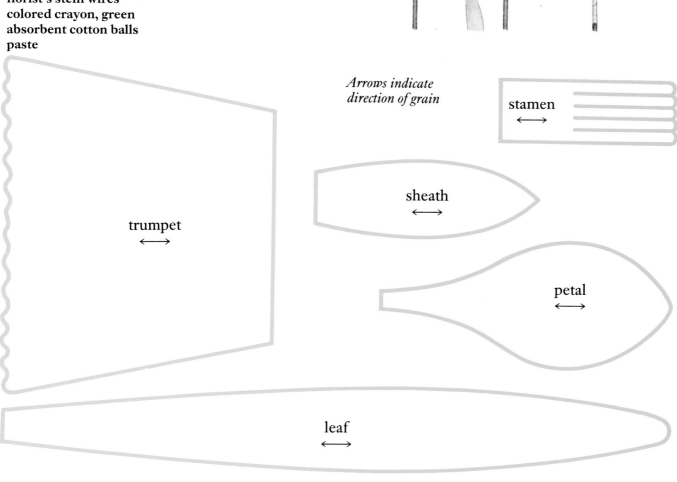

Arrows indicate direction of grain

stamen

trumpet

sheath

petal

leaf

To make the stamens, cut a strip of yellow crepe paper 1¾ inches wide. Along one side make cuts ⅛ inch apart and 1 inch deep. Snip the tops of the narrow strips to round off the corners. Cut the strips into sections, each with five stamen divisions. Using fine wire, bind one stamen section carefully to the wrapped wire, just below the base of the loop.

Treating the petals

To achieve a natural effect, slightly curl the top edges of each petal. Rest the back of a knife blade on the underside of the petal tip and, with your thumb on top, draw the blade up and outward so that the paper springs into a curl. Lightly color the base of the petals pale green. Rub green crayon onto the absorbent cotton and brush it gently over the petals, building up the color gradually.

The trumpet

To make the trumpet, flute the uneven edge of the shaped material. Hold the edge of the paper between forefingers and thumbs of both hands close together. Stretch the paper by pulling forward with the left hand and pushing away with the right hand, taking care not to tear the paper. About 1 inch from the base

of the trumpet strip, and with the inside facing you, gently stretch the paper between your fingers and thumbs so that it bulges and will form a cup shape. Wrap the trumpet to form a tube, overlapping the straight edge slightly. Paste the edge.

To give the beige sheath the look of dried plant material, gently crush the paper shape, then smooth it out.

Assembling the flower

To assemble each flower, place the trumpet over the stamen center and gather it tightly, just below the bulge. Secure it with fine wire so that the stamens are ½ inch below the fluted edge.

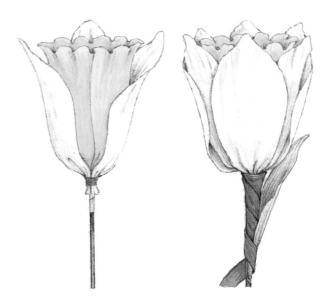

Bind the petals on about ¾ inch from the base, just above the point where the base narrows. Space the first three petals evenly around the trumpet and bind them securely. Then bind the remaining three in a second row, spacing them between the petals of the first layer. Make sure that all petals are at the same height. Snip off any surplus paper just below the fastening to neaten the base.

Covering the stem

Cut a ¾-inch strip of green crepe paper to wrap the stem. Bind the stem with the paper strip as described for the tulip (page 103). Wrap the stem with a second paper strip, adding the sheath 1½ inches below the flower. Using the fine wire, bind the leaves on opposite sides of the stem toward the base, then continue wrapping the stem evenly with the paper strip.

LASTING SUMMER ROSES

A few of these flowers and buds in a pottery basket or in a glistening crystal rose bowl, entwined around bare twigs in a flower pot or pinned to a looped "ribbon" of greenery to decorate a table, will give you lasting pleasure.

Materials you will need
(for 12 flowers or buds with leaf sprays)
cardboard, for petal and leaf templates
1 package heavy craft fabric, apricot (or chosen petal
  color)
1 package heavy craft fabric, leaf green
1 package crepe paper, leaf green
roll of fine wire, for binding
12 florist's stem wires, 8 inches long
steel knitting needle
paste

Cutting the templates
Trace the patterns of the petal, calyx and leaf onto cardboard and cut out the templates.

Making the petals
Place the petal template straight on the grain of the apricot craft fabric, draw around it and cut 12 petals

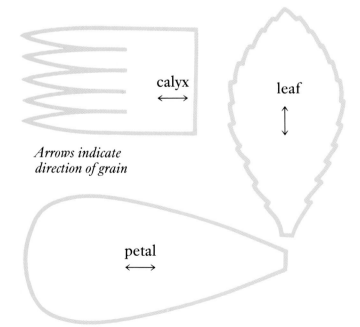

*Arrows indicate
direction of grain*

for each flower. Divide the petals into three groups.
Group 1: Stretch two petals deeply near the top, by holding each side of the petal with both thumbs and forefingers – this is called "cupping." Gradually stretch the entire width of the petal, carefully and evenly, into shape.
Group 2: Curl the top edges of six petals over a knitting needle. Cup them slightly in the center, curled top facing away from you.

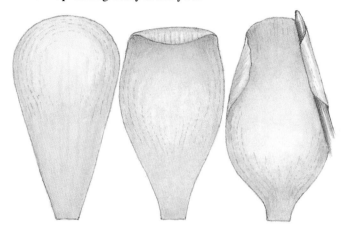

Group 3: Curl the remaining four petals deeply at the upper sides and cup them two-thirds of the way down.

Cut one calyx for each flower from the green craft fabric. Roll each tip into a point between thumb and forefinger.

Assembling the flower
To assemble the flower, take one of the wires and bend the top over about 1 inch to form a loop. Take one petal from Group 1 and roll it closely around the loop with the base of the petal ¾ inch below the loop. Secure the petal tightly with a twist of the fine wire just below the loop. Arrange the other petal around it, overlapping about halfway, so that the wire is hidden from view.

Arrange the six petals of Group 2 evenly around the first pair and bind them on in a similar way, with the tops of the petals a little higher so that the center will be slightly recessed.

Arrange the four petals of Group 3 at intervals around the others. Bind them in place and cut off the binding wire. Ease these petals outward to make a slightly open flower. Trim off the surplus paper beneath the flower. The bulk below the petals will form the calyx.

Brush paste around the base of the flower and adjust the calyx.

Making the leaves and stem

Cut the leaves from the green crepe paper. Make one or two leaf sprays for each flower. Each leaf spray consists of one large leaf, cut using the template, and two slightly smaller leaves.

To make the leaf sprays, cut a piece of fine wire 4½ inches long for each large leaf and 3 inches long for each smaller leaf. Bind the wire with a ¼-inch-wide strip of green crepe paper. Paste the wires to the center backs of the leaves, so that they reach about halfway along the leaves. Wrap the wire "stem" of the long leaf with another strip of the green crepe paper, binding in the smaller leaves on opposite sides of the stem about 1 inch below the large leaf. Mark veins on the leaves with a steel knitting needle, held like a pencil.

Bind the main flower stem of the rose with a ¾-inch-strip of green crepe paper as described for the tulip (see page 103). Wrap the stem with a second paper strip, binding in the leaf spray or sprays, one on each side of the stem about 2-3 inches below the flower.

Making the buds

To make a rosebud, cut two petals, one calyx, and one leaf spray if you wish. Also cut one 4-inch square of crepe paper in the petal color.

Loop a stem wire as for the rose. Double the paper square diagonally and roll it around the loop into a cone shape, overlapping and pinching it together ¾ inch from the bottom. Tightly bind this cone onto the stem. Trim off surplus paper.

Cup one petal deeply near the top. Curl the upper sides of the other over a knitting needle and cup deeply near the middle. Roll the first petal around the cone, then place the curled petal around it, overlapping it about halfway. Secure the petals with binding wire. Make a calyx as for the flower. Brush the base of the bud with paste and secure the calyx. Wrap the stem with a ¾-inch strip of green crepe paper, adding a leaf spray or sprays if desired.

A MASS OF COLOR

Making paper flowers allows you to acquire blossoms that span the seasons and are all shapes and sizes. The colors of your "instant" plants could complement each other and be chosen for one glorious arrangement. Here we give details for making anemones, bluebells, fuchsias and spider chrysanthemums.

Materials for all species
cardboard, for templates (for four species)
1 package crepe paper, mid green
roll of thick green wire, for stems
roll of fine wire, for binding
wire cutters
paste

ANEMONE

Materials you will need

1 package crepe paper, in chosen petal color(s)
1 package crepe paper, black
household bleach (optional)

Cutting and using the templates
Trace the patterns of the petal and the sepals, transfer them onto cardboard and cut out templates.

Cut a strip 2½ inches wide through a pack of crepe paper in your chosen petal color. Moving the petal template along, cut out the petal shapes right across this strip, through all the layers of the folded paper.

If you wish, dip the base of the strip in a weak solution of bleach and leave to dry. Open out the strip and cut off eight petals for each flower.

Making the flower centers and stamens
For each flower, cut a 9-inch length of green wire. Cut a 1-inch strip through all layers of the black paper. Open out the strip and cut off 6 inches. Then screw this into a flattened ball shape to cover the top of the wire. Cover this with a small length of the black strip, stretched across the center and gathered into the top of the stem. Glue in place.

Cut a 1½-inch strip through the black paper and make narrow, regular cuts along it. Open out the strip and cut off 4 inches. Tightly twist each cut section along the strip to make stamens. Glue around the flower center, and trim the stamens.

Assembling the flower

Gather the petals around the flower center and paste in place. From the mid-green paper, cut a strip 1½ inches wide. Moving the sepals template along (or cut out freehand if you prefer), cut out the sepal shapes right across this strip, through all layers. Open out, and cut off two 4-inch sections for each flower. Gather two sepal strips around the flower base and glue in place.

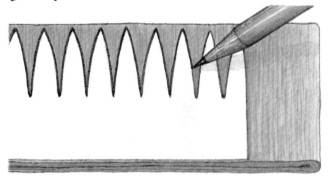

Covering the stem

Cut a 1-inch strip through the green pack. Use a length of this to bind the base of the flower and 1 inch down the stem. Twist another 4-inch sepal section carefully around the stem 1 inch down. Bind the full length of the stem twice with the green strip and secure with a little paste.

BLUEBELL

Materials you will need

1 package crepe paper, blue, or chosen petal color(s)
1 fine knitting needle

Cutting and using the templates

Trace the pattern of the bluebell heads, transfer onto cardboard and cut out a template.

From the blue pack, cut a strip 1½ inches wide and, moving the template along the strip, cut the bluebell pattern through all layers. Open out the strip and cut off six petals for each bell.

Making the stem and buds

For each stem, cut a 12-inch length of green wire. Cut a 1-inch-wide strip through all layers of the mid-green crepe paper, and through all of the blue or chosen petal color.

Open out the blue strip, cut off a 4-inch length, fold it in half lengthwise and shape it to make a small bud. Paste it to the top of the stem. Make another smaller bud and paste this bud beside the first.

Open out the green strip, cut off about 18 inches, glue the end to the top of the stem and tightly bind the buds and stem along the whole length as you build up the design.

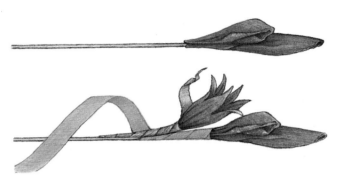

Making and attaching the flowers

Cut a 2-inch length of fine wire. Stretch the petal strip to cup it along its length, roll it around a finger, slip the bell shape over the wire and bind it tightly with fine wire.

Cut a ½-inch strip through the mid-green pack and cut off enough to bind the bell and stalklet as tightly as possible. Curl each petal by rolling the top over a fine knitting needle.

Make five more bells in the same way. Add the bell flowers one by one to the main stem at regular intervals, binding them in place with the green paper strip. Bind to the bottom of the stem, then bind it again with a second paper strip and secure the end with a dab of glue. Gently bend the top of the main flower stem into a curve.

Making and attaching the leaves

To make the leaves, cut a 9-inch length of green wire. Glue it between two layers of green paper and cut out a narrow spear shape. Gather the paper at the base and glue in place on the stem.

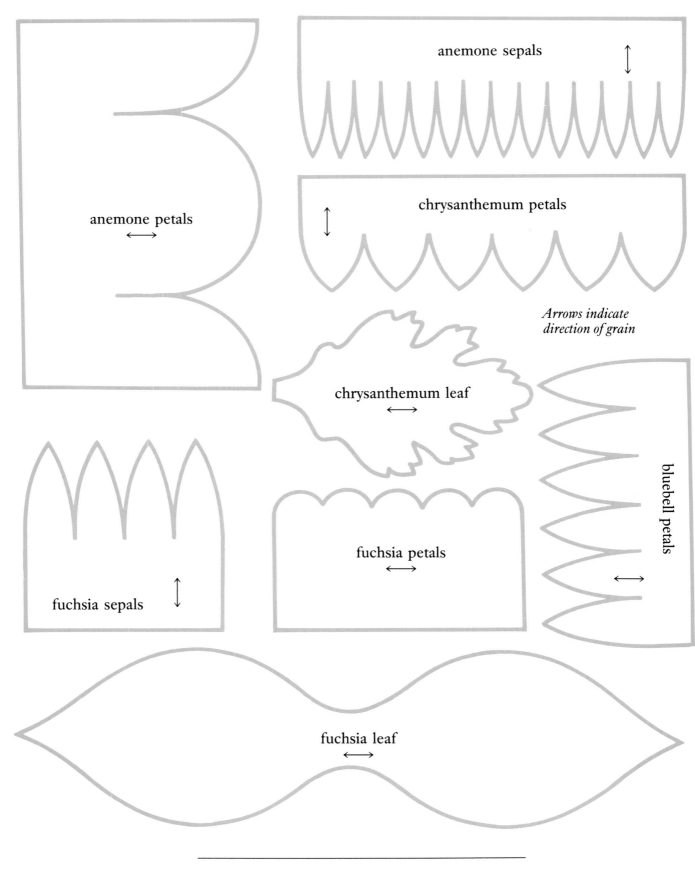

anemone sepals

anemone petals

chrysanthemum petals

*Arrows indicate
direction of grain*

chrysanthemum leaf

bluebell petals

fuchsia sepals

fuchsia petals

fuchsia leaf

FUCHSIA

Materials you will need
1 package crepe paper, white, for petals
1 package crepe paper, scarlet, for sepals
1 package crepe paper, light brown
1 fine knitting needle

Cutting the templates
Trace the pattern of the leaf, petals and sepals, transfer onto cardboard and cut out templates.

Making the flowers
Cut strips 1½ inches wide, ½ inch wide and 2 inches wide through white pack. Moving the petal template along the 1½-inch strip, cut out the shape through all layers. Open out the strip and cut off four petal sections for each flower. Twist one length and cup it to form a bell shape. Bind the bell-shaped base with fine wire, leaving a short end to attach the flower to the branch.

From the contrasting color, say scarlet, cut strips 2 inches wide and ½ inch wide. Moving the sepals template along the 2-inch strip, cut out the shape. Open out the strip and, for each flower, cut off one four-sepal section. Gather the strip around the base of the flower and glue in place.

Open out the narrow contrast strip and use a short length to bind over and over the ends of the sepals. Open out the narrow white strip and use a short length to cover the seam and bind the flower stalk.

From the 2-inch-wide strip of the main color (white), make regular narrow cuts through all layers. Open out the strip and cut it into sections of six stamens for each flower. Twist the stamens tightly, roll up the strip, dab glue around the base and push it well down into the flower center. Cut five stamens short and bend the end of the sixth.

Making the buds
To make a bud, cut a 5-inch length from the 2-inch-wide strip of the main color (white). Shape it to form a bud and bind the base with fine wire, leaving a short end to attach the bud to the branch. Add sepals in the contrast color, and bind and finish as for the flower. Bring the four points of the sepals together and glue at tip.

Making the leaves
To make the leaves, cut out the pattern three times through double thicknesses of mid-green paper. Cut three 4-inch lengths of green wire and glue one length down the center of each leaf. Press the two thicknesses together while the glue is wet to make double leaves.

Making the stem
For each stem, cut a 14-inch length of green wire. Cut a 1-inch-wide strip through all thicknesses of the light brown crepe paper. Paste one end of this strip to the top of the wire and then bind the stem, stretching the strip tightly. Cut off and secure the end with glue.

Assembling the flower branch
Twist a double leaf shape to the top of the main stem. Bend over the brown stem at the top to hold it firmly. Add three flower stalks to this point at the top of the branch, using another length of the light brown paper strip, and glue them in place. Working down the branch, add another pair of leaves. Bind and glue in place. Continue down the branch, adding another pair of leaves. Secure the binding at the base of the branch with a dab of paste. Curl back all the sepals.

SPIDER CHRYSANTHEMUMS

Materials you will need
1 package crepe paper, in chosen petal color
1 package crepe paper, dark green
pinking shears
household bleach

Cutting the templates
Trace the patterns of the sepals and the leaf, transfer them onto cardboard and cut out templates.

Making the flowers
Cut a 3-inch-wide strip through a pack of paper in the petal color. Cut into the strip at regular intervals through all layers of the paper.

Open out the strip to double the packet width, about 8 inches, and trim down about ½ inch with pinking shears. Do *not* cut this off. Continue to open out the strip to four packet widths – about 16 inches – and then cut off from the remainder. Twist each individual petal carefully between finger and thumb to make spikes.

Take the pinked edge of the strip and begin to roll it up, pinching and gathering as tightly as possible. Roll up the rest of the spiky petals.

Cut a 12-inch length of green wire and twist the top around the base of the flower head, pulling the stem downward. Trim off excess paper at the base of the flower head. Dip the ends of flower spikes in weak bleach to lighten the tips.

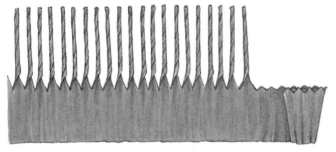

Making the sepals
From the dark green pack of paper, cut strips 1¼ inches wide and ½ inch wide through all layers. Moving the sepals template along the 1¼-inch strip, cut out the pattern. Open out the strip and cut off an eight-sepal section for each flower head. Gather the sepal strip around the base of the flower head and glue it in place. Fan out the sepals and glue four of them to the underside of the flower head, making it fairly flat.

Covering the stem
Paste the top of the ½-inch green strip to the top of the stem and bind it down about 1½ inches. At this point make a loop of paper (to form a leaflet later) and continue to bind the stem for another inch. Cut off the end and paste the binding down.

Making the leaves
To make a leaf, use the template to cut the pattern through two layers of the dark green paper. Glue and bind to the stem with another length of the dark green strip. Continue to the bottom, cut off and secure with glue. Shape out the loop to make a small spear-shaped leaflet. Shape and refine the main leaf. Mark veins with the point of a knitting needle.

A GIFT OF GRAPES

There are so many ways to display these bunches of silky grapes that you will be tempted to make a bumper harvest. Turn a bottle of wine into a vintage gift by decorating it with appropriately pale green or red grapes and a clinging tendril of vine leaves.

Materials you will need
(for one bunch of 24 grapes)
cardboard for leaf template
piece of fine, grape-colored fabric, silk or cotton lawn, about 30 inches square
small piece of coarse dark green fabric, such as flock material, about 10 inches square
roll of medium-gauge wire, to outline leaves
roll of fine wire, for binding
1 florist's stem wire
florist's binding tape
absorbent cotton balls
glue

Making the grapes
Cut 1¼-inch squares of the fine material for the grapes. Tear off 24 small pieces of absorbent cotton, roll each one in a square of the fabric and bind with a length of fine wire, leaving about 2 inches trailing.

Cover one end of the stem wire with binding tape, wrapping it tightly to conceal the stem. Add one of the grapes to extend beyond the wire end, and bind the trailing wire around the stem. Add two more grapes close against the stem, twisting the fine wire ends around the stem and covering them with florist's tape as you work down the length. Add three more grapes in the same way, and continue until they are all in place. Make sure that the bunch fills out neatly from the tip to make a full cone shape.

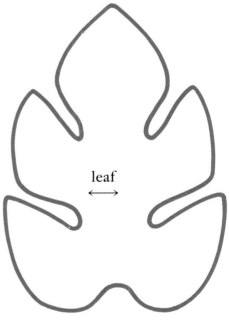

leaf
←→

Making and attaching the leaves

Trace the pattern of the leaf onto cardboard and cut out the template. To make one leaf, fold the green material in half and cut out the template shape through both thicknesses. Stick the medium-gauge wire along the length of one leaf shape, leaving about 2 inches trailing for the stem. Glue the second leaf shape on top of the first leaf shape.

Twist the leaf wires around the stem near the top and bind to the end of the stem with the tape. Finish off the end neatly.

Ease the grapes into a neat shape and twist the leaves so that they look realistic.

RIBBON ROSES

A few strips of ribbon and a few moments to spare are all you need to make these dainty ribbon roses. They are a good stepping-stone for readers who are new to the craft of making artificial flowers – there are no patterns to cut and very few materials are needed.

Make the ribbon roses to trim a pretty package, and they become a lasting keepsake. Make a spray of the flowers in pink, red, cream, yellow or white, and wear them on a lapel or a summery hat, and they are a flattering fashion accessory, or mount the flowers on long stems and use them to decorate a bedroom.

LARGE ROSES

Materials you will need
(for one flower)
35 inches double-sided satin ribbon
  (width of the ribbon may vary from 1½–3 inches
  according to size of flower required)
florist's stem wire
roll of fine wire, for binding
florist's binding tape, for stem

Making the flower
Bend the stem wire into a narrow loop at one end, making a hairpin shape that is about 4 inches in length.

Starting from one end of the strip of ribbon, fold it over at a right angle. Continue folding from each side in turn, folding toward you and forming the ribbon

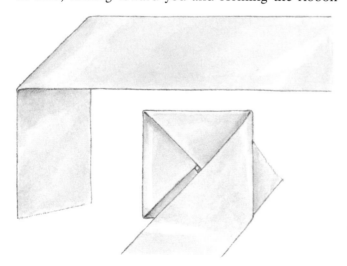

into a square with a small hole in the center. Hold the ribbon square firmly as you work, or it will spring up and out of shape. Continue until only 2 inches of ribbon is left unfolded.

Hold the ribbon square firmly in one hand and push the looped wire up through the center opening. Thread the remaining 2-inch strip of ribbon through the loop and twist it clockwise so that the ribbon coils around and around the loop and forms a realistically roselike center. Pull the wire gently until it is concealed in this center coil.

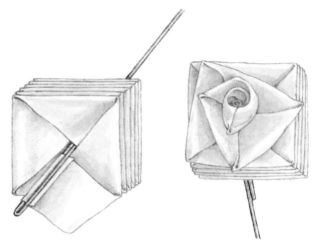

Take the first fold underneath the flower and bind it securely to the stem with fine wire.

Bind the base of the rose and the length of the stem with florist's binding tape.

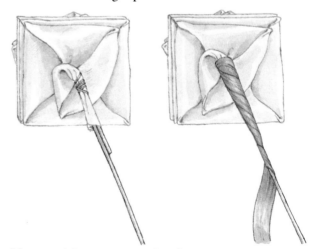

If you wish to mount the flower on a long stem, push a length of thicker wire up into the center of the flower, bind it to the stem wire with fine wire, then cover it all neatly with the florist's tape.

SMALL ROSES

Materials you will need
(for six flowers)
72 inches satin ribbon, 1 inch wide
18 inches medium string, for stems
needle and thread
12 small rose leaves, from hobby shop
florist's binding tape, green

Making the flower
To make each flower, cut 12 inches of ribbon and 3 inches of string. Make a knot close to one end of the string.

Twist one end of the ribbon around the knot, fold over the corners and stitch the ribbon to enclose the knot. Tack 4 inches along one edge of the ribbon with running stitch and pull up the stitches to gather the ribbon.

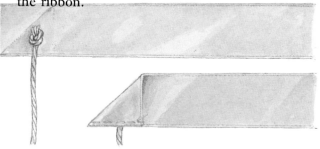

Wrap the gathered ribbon around the flower center (the covered knot) and stitch in place.

Tack along the next 4 inches of ribbon, pull up the thread, wrap it around the flower center and stitch in place. Repeat with the last 4 inches of the ribbon, sewing it in place.

Wrap florist's tape 10 times around the base of the flower to form the calyx, making sure to cover the base of the ribbon. Wrap the tape for another ½ inch,

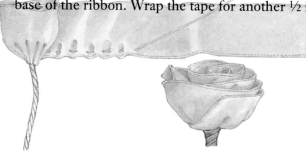

insert the leaf wire and continue taping to the base of the string stem.

SUNFLOWERS

Made from ordinary upholstery-quality vinyl and bright and brazen as can be, these flowers make a bold splash of color with all the impact of a field of sunflowers on a golden summer's day.

Materials you will need
(for six flowers)
cardboard, for petal template
40 inches stiff vinyl material,
    25 inches wide
1-inch-diameter buttons
small piece of brown fabric, to cover centers
roll of fine wire
6 florist's stem wires
roll of medium-gauge wire, for stamens
roll of thick wire, for stems
florist's binding tape, green
brown poster paint
absorbent cotton balls
glue

Cutting the templates
Trace the petal pattern onto cardboard and cut out the template.

Use the template to cut 24 petal shapes for each flower – that is for 12 double-thickness petals.

Making the flowers
Cut 6-inch lengths of fine wire. Glue a piece of wire down the centers of half the petals so that the wire extends 2 inches below the petal. Glue the other 12 shapes on top so that the wire is enclosed in double-thickness petals.

Push one end of a florist's wire up through one hole in the button and down through another. Twist the ends underneath the button.

For each flower, cut a 3-inch square of brown material. Cover the button with a small piece of absorbent cotton and then with the brown fabric. Bind the fabric tightly under the button with fine wire.

For each flower, cut 2½-inch lengths of fine wire to make the stamens. Twist a very small piece of absorbent cotton tightly around one end of each length and cover it with green binding tape, wrapped around very tightly and finished off neatly. Dab the stamen ends with brown poster paint and leave to dry

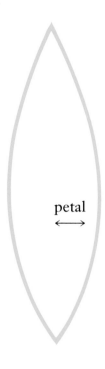

petal
←→

before assembling the flower center.

Arrange the stamens around the button representing the flower center, spacing them evenly.

Bind the stamens beneath the button with fine wire.

Arrange the petals evenly around the stamens and flower center and bind each one in place.

For each flower, cut 20 inches of the thick wire. Push the wire up through the ring of wired-on petals until it reaches the underside of the button. Twist the exposed length of stem wire around the thicker wire.

Starting at the base of the flower, bind the stem tightly and neatly with the green binding tape. Press the plastic forming the base of the petals tightly to form a neat calyx shape as you work. Continue binding to the end of the thick stem. Cut off the tape and bind the stem over again. Make the other five flowers in the same way.

CHRISTMASTIME

A stocking trimmed with a brilliant poinsettia, evergreens and baubles will be a delightful surprise for Christmas, but do not give this decoration to a very young child as the wires and stamens could be dangerous.

Materials you will need
(for one poinsettia)
cardboard, for petal template
30 inches red flock ribbon,
   2¾ inches wide
10-inch square green felt
6 red artificial stamens
5 yellow artificial stamens
florist's stem wire
roll of fine wire
florist's green binding tape, to bind stems
1 absorbent cotton ball
glue

Cutting the templates
Trace the pattern of the two petal sizes onto cardboard and cut out the templates. Fold the ribbon in half lengthwise. Use the templates to cut three large and nine small pairs of petals from the double thickness of the ribbon. Make the holly leaf template and cut five pairs of leaves from green felt.

Making the flowers
Glue a length of fine wire down the center of one side of each petal, dab glue around the edge and stick down the other petal halves, so that you have double-thickness petals with wire in the center. Set aside to dry.

Bind the four lengths of stem wires together with fine wire to form the stem. Arrange the stamens around the top and bind them on with wire.

Bind the nine small petals at equal intervals around the stem one at a time, evenly spaced around the

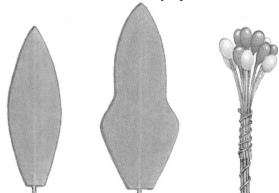

inner ring. Finally, bind on each of the three large petals, spacing them equally around the others.

Wrap a small piece of absorbent cotton around the base of the flower, to represent the calyx. Bind from the base of the flower, over the calyx and down the length of the stem with florist's wire.

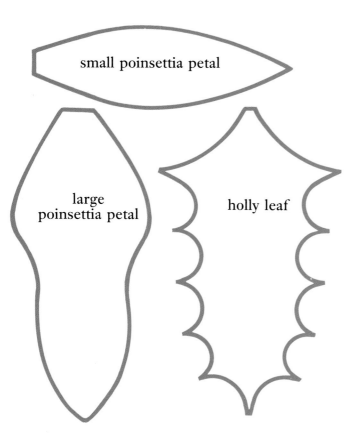

small poinsettia petal

large poinsettia petal

holly leaf

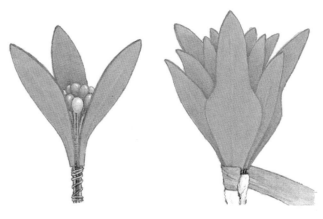

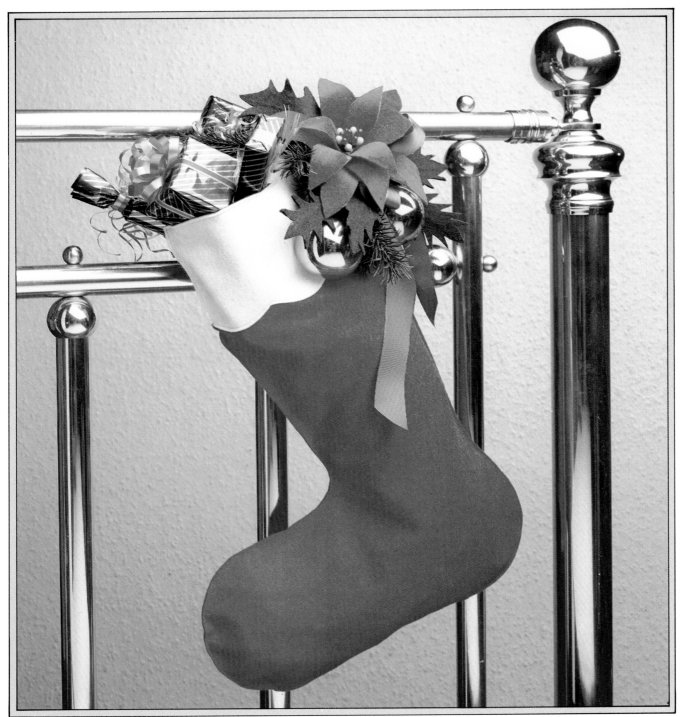

Making the leaves
Glue a length of stem wire down the center of one leaf from each pair, then dab glue around the edge of each wired leaf. Place a matching leaf over each glued leaf and bind the stem with florist's tape. Twist the leaf stems around the poinsettia stem.

To assemble the decoration, make a fan shape of the holly leaves, adding fresh sprigs of evergreen if you choose, around the poinsettia. Sew the poinsettia decoration and baubles to the top of the stocking.

SIMPLY ELEGANT

There is more than a hint of the twenties and thirties in these elegant nylon flowers. Pale, romantic shades of pink, lilac and cream combine in a way that sets the imagination racing.

Display a bunch of these beautiful blooms in a classic urn-shaped vase to stand on a pedestal, arrange them in a long, low container to grace the center of a dining table, cluster a few blossoms in a pretty china basket under a lamp or on a dressing table, or attach them to a picture or mirror frame.

Materials you will need
(for six flowers)

very fine stretch material, such as nylon stockings or pantyhose, georgette or chiffon, in petal colors, and in green, for leaves

bunch of artificial stamens with colored, silver or golden tips

roll of fine wire, for binding

copper, galvanized or millinery wire to outline the edges of the petals (the wire must be easy to manipulate)

florist's stem wires

florist's binding tape, green

Making the flowers and leaves
Arrange five or six stamens around one end of a stem wire and bind them tightly in place with fine wire. From the wire used for the petal and leaf shapes, cut lengths of about 5 inches for each petal and slightly shorter lengths for the leaves. Bend each piece of wire into a loop and twist the ends to secure them, leaving some length to attach to the stem.

Push the wire loop into the fabric and stretch it tightly. Draw the fabric into the base of the loop, bind it tightly with fine wire and cut off the surplus.

Tightly bind each petal onto the stem wire and stamens group, making sure that it is firmly attached and will not slip.

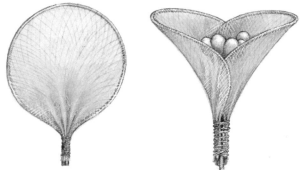

Start binding the flower high up on the stem, where the calyx would be, making sure that all the binding wires are completely concealed. Bind tape down the full length of the stem wire stem, cut off the tape, then bind the calyx and stem again.

Make leaves in the same way as petals, using fine green stretch material. Bind the leaves onto the stem with fine wire and cover the seams neatly with tape.

You will find it surprisingly easy to shape the petals, twisting and bending the wire outlines. This final shaping of the flowers is the most important stage, giving each flower a distinctive form.

Sprays of small, tight buds are a useful feature in a group of these dainty flowers, especially if you are arranging them in a shape – such as a long, low horizontal or a triangle – that benefits from having clearly defined points.

Make the buds in the way described for the grapes (page 116), spacing them at intervals on each side of the stem, in pairs, not quite opposite each other.

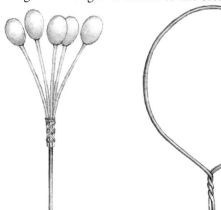

READY-MADE FLOWERS

Ready-made flowers can be all things to all people. You can buy some that are so convincing, visitors just cannot resist touching them to see if they are real. There are also improbable blooms, that are obviously fakes and in frankly fantastic or dazzlingly outrageous colors. Our purchases from department stores include both.

Those artificial flowers that resemble garden varieties look especially pretty arranged with preserved leaves or long-lasting evergreens. Often the leaves on sprays of artificial plant material do not reach the same high standard of texture, shape and color that is achieved with the flowers, and are disappointing in an otherwise aesthetically pleasing group. Strip them off, add natural foliage, and the arrangement becomes softer, prettier and more believable.

Containers for realistic artificial flowers can be chosen without regard to the flowers' pedigree – or lack of it. Choose any holder that would complement fresh or dried flowers. Silky sweet peas might tumble from a pretty china teapot; you could pile roses into a glistening crystal bowl; allow freesias and bluebells to form gentle, natural-looking curves in a pedestal vase or fill a shallow basket with poppies or peonies.

BUILT-IN ADVANTAGES

Ready-made flowers with wire stems will stand upright, but use a strong pair of wire cutters or pincers from the tool box if you want to cut such flowers as you arrange them; or have a pair of pliers on hand if you decide to bend back the stems to shorten them. This way you can unravel the stems to their original full length when rearranging the flowers later. Shortened wire stems can be extended another time by twisting them on a florist's stem wire.

Anchor wire stems in a holder – useful for an arrangement in a shallow dish or on a wooden board – in dry, brown stem-holding foam or a handful of glass marbles.

The stems of the peaches and cream-colored flowers arranged in a basket (see title page) are held in a block of foam that is tied securely, parcel fashion, with fine twine taken through the basket base.

Arrange the leaves first to give the triangular, curving or casual outline you have chosen. Glycerin-preserved leaves such as beech, lime and oak blend particularly well with artificial flowers, the curving branches giving a graceful look to the design, their suppleness and sheen perfectly complementing the silky petals. For the "long points," the natural shape contrast to full, round flowers, choose preserved stems of rosemary (which retains its aroma), berberis and yew. And for the visual weight at the base of a design, arrange a cluster of single leaves in preference to one or two very large ones – such as preserved maple, magnolia, camellia or geranium. Bleached preserved leaves, in all the deep, buttery-cream tones, are preferable to those in very dark shades. Large, dark-colored foliage can dominate delicate colors and "fragile" flowers, and artificial ones are no exception.

FANTASY HARVEST

For a complete contrast to the pastel shades of the flower-basket design, we shopped around and produced an improbable harvest of poppies, peonies and lilies all in black and gray – with just one colorful highlight, a glowing red poppy.

Contrast in texture and good lighting are the secrets to displaying them with style. Glittering, shiny containers are just the thing. Polished silver, brass and copper, chromium-plated and tin items catch every glint of light and add sparkle to the group. Chunky glass containers are good too, providing just the right textural contrast. Conceal a piece of dry stem-holding foam in a heavy glass vase, fill it with shiny marbles and give the darkest of artificial flowers a good send-off.

Consider the background carefully. These flowers look best against the shiniest of surfaces such as gloss paint, metal window blinds, plastic-coated fabrics and mirror glass. These are the surfaces that attract and hold the light, reflecting it back onto the flowers – and those flowers in deep, midnight shades need all the limelight they can get. Place them in the full glare of the sun, so that every shaft of light from the window reaches deep into the centers and brings out the detail in the petals; arrange them in directional light from a wall or ceiling spotlight or table lamp, or on a glass surface that is lit from below.

Store-bought poppies, peonies and lilies will make an effective feature held with heavy marbles in a large glass container.

TREES FOR ALL SEASONS

The brilliant colors and varied shapes of artificial flowers – as with natural ones – inspire a forest of ideas for lasting decorations in the home, and even the backyard. As our designs on the preceding pages show, the possibilities extend far beyond arranging the flowers in a basket or vase.

One pretty way to use artificial flowers is to arrange them into clipped tree shapes, making your own imaginative pieces of topiary. Because artificial flowers are so long lasting, it is worth spending a little time arranging them into pleasing designs.

You can make simulated flowering trees on any scale you choose – miniature ones set in egg cups, to stand at each place around the dining table; slightly larger ones in elegant pottery planters for a table centerpiece; all the way to tall and graceful blossoming trees in earthenware flowerpots or wooden tubs to stand in the corner of a room or on the patio or balcony.

Designer trees may take the form of a formal shape atop the trunk, such as a neat sphere or regular cone, with the flowers and foliage placed to keep within a clearly defined outline, or be more disheveled in appearance. Only a generation or so ago, such decorations were made using dried apples and oranges – themselves neat spheres – or clumps of modeling clay to hold the flower stems. Now the creative task is made much simpler by the use of pre-formed shapes of nonabsorbent plastic foam, especially made for use with artificial flowers and dried plant materials, or water-holding foam, which is used for designs of fresh flowers and foliage.

The spheres range from golf ball size to volleyball size (measuring about 9 inches in diameter) and the cones are made in an almost equally wide range. By selecting flat, "single" flowers and securing them close to the holding material, you can keep your design within tight size limitations – little larger indeed, than the holding shape itself. If you insert flowers and foliage on longer stems and add knobby or spindly twigs for good measure, you can greatly increase the overall size of the decoration, perhaps trebling the circumference. The foam shapes are available from most florists or florist supply shops.

Other ways of holding the stems can also be devised.

This small, formal tree is composed on a cone of nonabsorbent foam mounted on an apple twig. The silken "flowers" represent cream and white fruit blossoms and deep pink primula. Two ribbon bows with trailing ends complete the festive design.

You can shape a ball of pliable modeling clay around the tree trunk, or use a clump of damp plaster of paris – though be sure to insert the flower stems quickly, before it dries out and sets solid. Another useful stem-holding device is a piece of wire netting tightly crushed to form any shape, in any size you wish. Choose small or large holes, according to the scale of your design, and mold it around scraps of dry foam or dry sphagnum moss to give the stems extra stability when they are pushed through the criss-cross of wires. (If you wish to use fresh flower materials for a tree, use soaked foam or damp moss to provide a constant source of moisture.)

A sense of security is essential as you build up the design at every stage. First you must anchor the tree trunk firmly in the container, then make sure that the stem-holding material cannot slide down the trunk and lastly, of course, that the flowers are securely attached.

Many-colored cosmos and freesias give this tree a country garden look. The holding material, mounted on a bamboo cane, is a ball of dry foam covered with clumps of sphagnum moss.

If the container is to be a flowerpot, with a central hole for drainage, this will help you to locate the trunk, and for a small design, a ball of modeling clay or extra-tacky florist's clay pressed firmly around the base and the vertical stick will probably be enough. For larger trees, heavy stones and pieces of brick wedged around the trunk help – the weight at the base is useful to avoid toppling – and for the ultimate in rigidity, pack plaster of paris around the rubble. If you first line the pot with thin slices of dry foam, it will eliminate the danger of cracking the pot and will enable you to lift out the solid block of plaster when

eventually you want to dismantle the design.

Fill up the pot with dry peat moss, sand, potting soil, foam chips or gravel. For a pretty finishing touch, you can scatter a handful of colorful potpourri (dried flower petals) on the top.

Once the trunk is firmly secured, attach the stem-holding block securely to it. To do this, it may be necessary to cut a notch in the wood just below the foam block, and bind thick wire tightly around it. A ribbon bow tied around the trunk will hide the binding.

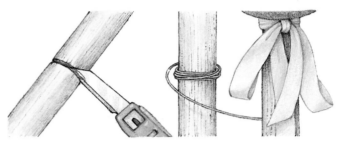

If you are creating a close-cropped design, with the flowers resting in, say, concentric circles or a spiral pattern close against the shape, you will have few problems in concealing the inner mechanics – perhaps just a few gaps, which can be filled in with small flowers or leaves. For designs that use long-stemmed flowers and twigs, it is best to cover the holding material completely first. Pads of dry sphagnum moss or lichen – which you can secure with hairpins or staples of twisted wire – or short sprays of flowers such as dried *Alchemilla mollis* or hydrangea, will conceal the base.

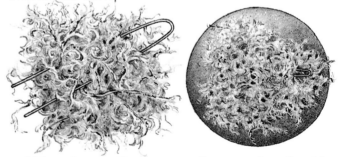

Although you will select your flowers primarily with an eye for their color, texture and shape – and for their weather resistance, if the tree is to be placed outdoors – also check the strength and rigidity of the stems. Most artificial flowers, even the delicate-looking, silky ones, have firm plastic- or fabric-covered wire stems that may be pushed easily into foam or holding clay. Some types, however, have soft, flexible stems, and every single one of these would need to be wired (see page 61) before being arranged.

Chapter 4
HEALTH AND BEAUTY

For thousands of years flowers have formed a hallowed part of our cultural heritage. They have been grown for the beauty care of queens – kings, too – and regarded as symbols of purity. Their fragrance has been used for personal enhancement and, in the form of aromatic oils, has had deep significance in religious ceremonies. The wonderful colors that can be extracted from plants have for centuries been used to dye yarns of wool, cotton and silk, for weaving into clothes, rugs and other furnishings.

The earliest use of flowers and other plant materials in all these ways was, of course, making a virtue out of necessity. People were entirely dependent on the natural products to be found locally or, in the case of spices, what could be easily obtained from imports. Today, when every fragrance and hue can be accurately reproduced in a laboratory, we can preserve our complexions, and immerse ourselves in tantalizing aromas, draping ourselves and our homes in brilliant colors, entirely without recourse to a single natural product.

But "beauty from a bottle" may not match the personal satisfaction of capturing the aroma and color of the natural products from the world around us. If you love nature and have reservations about the methods used for some laboratory and research experiments, and have even a little leisure time to spare, why not make your own beauty – and beautiful – products and share this delight with your friends.

EARLY COSMETICS

As befits any woman's beauty preparations, the exact composition of the first cosmetic products has been lost to time. What is clear is that these recipes were known in one form or another to most ancient peoples. In some cases the use of aromatic ointments and oils was largely to protect the skin and hair against the ravages of hot and dry climates; some preparations were used as status symbols; some had a religious significance; and others had magical associations, as in the case of warriors who smeared their bodies with pungent oils before going into battle, relying on the oil's mystical powers to bring about rapid victory.

The base of all the earliest cosmetics must have been natural vegetable oils and animal fats that were available. The first known cleansing cream, a mixture of olive oil, wax and rosewater, was blended by a

Homemade soaps, herbal oils and shampoos are an inexpensive and delightful alternative to commercial products.

Greek physician, Galen. All along the Mediterranean ladies came to value sesame oil as a skin food that penetrated and softened the skin and, since it absorbed most of the ultraviolet rays, protected the skin from the sun.

Athenian ladies liked to burn a little powdered myrrh on a stone, allowing the astringent resin fumes to penetrate the skin in the form of a fragrant, cleansing steam bath, and so keep wrinkles at bay.

Fashion played its part, and in every society there was a natural trendsetter. Cleopatra built up her confidence – and presumably weakened the political resolve of her adversaries – by appearing awe-inspiringly decorative at all times. History tells us that she used floral-scented face masks to soften and clarify her skin, and cosmetic creams colored with saffron. She rubbed fragrant oils into her body, painted her lips and cheeks with red ochre and accentuated her eyes with generous applications of kohl. Other beauty tricks of the time included the use of butter and barley flour paste scented with flower oils to eradicate skin blemishes (the Romans used heavily perfumed white lead and chalk for the same purpose) and henna, a deep, reddish-brown dye, to color the hair, nails and palms of the hands and soles of the feet. To this day, henna is regularly used in this way in North African countries. This author had a narrow escape from a friendly group of Bedouin tribeswomen in Tunisia who made every attempt to transform her appearance from top to toe with a pot of the omnipotent dye and a menacingly wide brush!

PETAL AND SPICE BATHS

The Egyptians considered frequent and fragrant baths a prerequisite to purity and good health, and in Syria and Rome it was the custom for kings and emperors to join forces with their loyal subjects in bathing pools. In Syria, on these occasions, the water was lavishly strewn with rose petals, a water-borne red carpet spread before the monarch. Roman emperors vied with each other to distribute publicly the most fragrant of oils. The emperor Heliogabalus, renowned for his personal extravagance, bathed in saffron-scented, golden-colored water, emerging with a glowing though temporary tan imparted by the powdered stigmas of blue-violet autumn crocus flowers.

One of the oldest known body lotions, Queen-of-Hungary water, is said to have won the approval of that particular royal lady not only as an after-bath refresher but because she believed it had cured her paralysis. A precise list of ingredients has not survived the passage of time, but flowering rosemary, grated spices, dried lemon rind and pure alcohol were among them.

Perfumed baths attained wide popularity in England in the 17th century, when women would gather flowers into bath bouquets to scent the water,

Cosmetics blended at home with natural herbs and oils can be made to suit your own preferences.

and splash themselves with flower-scented wine or vinegar. Recipes for these bouquets were as individual as the gardens where they grew. They would include some fragrant flowers such as rose, honeysuckle and lily, and herb leaves such as scented geranium and lemon balm. In those days the bouquets were tied with ribbons and trailed in the water. Today they might be hung beneath the bath faucet. Marie-Antoinette's chosen bath fragrance, which was made up into "muslin bonnets," consisted of equal measures of wild thyme flowers, rosemary leaves and sea salt.

Fresh petal and spice mixtures, the equivalent of dried potpourri blends, were widely popular as bath fragrances. Some would be sweetly scented, including rose petals, thyme and lavender flowers, crumbled bay leaves or mint tied into muslin bags; others would be more pungent – and carried by fashionable males – and include, besides rose petals,

marigold and lemon balm leaves, crushed juniper berries and fenugreek seeds. (Middle Eastern harem ladies, it is worth noting, preferred to eat their fenugreek seeds, in the belief that they were the secret of rounded plumpness of the body.) Perhaps the best bath aromatic of all – one you can achieve when, on a sudden whim, you feel like pampering yourself – is simply that brought about by scattering rose petals and a few drops of oil of roses into your bath water.

ASH AND LAVENDER

Soap is reputed to have been first discovered by accident in Rome about 1000 B.C., when the fat from sacrificial animals combined with ashes from the fire and set into blocks. Gradually spices and floral scents were added to color and perfume this soap, making its use as much a pleasure as a practicality. Examples of the soapmaker's craft have been found in excavations in Pompeii, Italy.

Lavender (the plant name derives from the Latin *lavare*, to wash) has had a long association with bathing and also with relaxing the mind and body. An old herbal book invites its readers to boil two handfuls of lavender flowers in water with a little salt, strain the infusion and put it into a bath "not too hot, in which to bathe the body in the morning and at the same time to cleare the minde."

Town dwellers who were denied the therapeutic luxury of stepping into the herb garden and cutting lavender flowers for the bathroom and boudoir could step outside their front doors and buy them. "Sweet-smelling lavender, six fine bunches for one penny" was one of the frequently heard flower cries of London and other cities. Roses, marigolds, myrtle, stocks, mignonette, rosemary and thyme were other flowers sold and "cried" on the streets.

FRAGRANT OILS

Natural oils impregnated with the aroma of flowers, leaves and seeds are the key to many cosmetic and beauty preparations, soaps, bath fragrances, hair lighteners and conditioners, skin-care creams, perfumed candles and traditional country recipes. As each season brings its harvest of aromatic plants, several fragrant oils can be made. You will soon build up a delightful collection to ring the personality changes on a range of beauty crafts.

Extracting the essential oils from plants is a job for specialists. The mistresses of the potpourri, on the great English estates, spent a lifetime perfecting their craft. But by infusing pure vegetable oils with fragrant plants you can make pleasing substitutes; or, for a more concentrated aroma, you can buy mini bottles of "the real thing" in herbalists' shops.

FLORAL OIL

To make floral oils first select your fragrance according to preference. Rose, carnation, hyacinth, jasmine, lilac, stock, lily and honeysuckle are but a few of the options.

1¼ cups oil (such as sunflower, safflower, almond, corn or olive)
10 large cupfuls petals
1 tsp. liquid storax
1 tsp. tincture benzoin

1. Pull off the petals of large flowers, and strip small petals from the stalks. You will need enough petals to pack into 10 large cups. Measure two cupfuls and store remaining petals in a plastic bag in the refrigerator.
2. Pour 1¼ cups of your chosen oil into the top of a double boiler or a bowl, over a pan of simmering water. Heat the oil, then add two cups of flowers or petals. Stir, cover tightly and leave on very low heat for two hours. Strain and keep the flowers. Add another two cups of flowers to the oil and repeat the process. Then repeat this three times more.
3. Pour the oil and all the flowers into a pan, bring it slowly to a boil, and simmer over low heat for about one hour, or until the flowers are dry. Strain the oil through muslin, pressing the dried flowers with a spoon against the strainer.
4. Stir in the liquid storax and the tincture of benzoin to fix the aroma. Bottle and cover the oil, label it and store in a dry, dark place.

HERBAL OIL

Making herbal oil is a simpler, but longer process. Applemint, peppermint, lemon balm, sweet cicely, rosemary and fennel seed oils can be kept on hand in the kitchen for culinary as well as beauty craft purposes, provided they are kept in a dry cupboard.

1¼ cups oil
3 tbsp. herbs
2 tbsp. wine or cider vinegar

1. Gather fresh herbs, strip the leaves from the stalks and crush or grind them until they are like coarse grains. For each 1¼ cups oil you will need 3 tbsp. herbs. Half fill a jar with the herbs and oil, and add 2 tbsp. wine or cider vinegar.
2. Cover and shake the jar and leave it on a sunny windowsill or over a radiator for three weeks. Shake the jar every day.
3. Strain the herbs, add 3 tbsp. fresh herbs and leave for a further three weeks. Do this once more, then finally strain the oil through muslin. Pour it into a clean bottle, cover and label it, and store in a dry, dark place.

FRAGRANT SOAPS

Decide to delve into the fascinating craft of soapmaking and, in the kitchen, you will become not a cook but a practicing chemist, chopping herbs, selecting bottles of fragrant oils and mixing everyday ingredients such as honey, oatmeal and vegetable oils.

Making "real" soap is a task that needs meticulous care because it involves using lye, a powerful alkali that can cause serious skin burns. It is important that you can work without distraction and be sure that there will be no flying leaps onto the table from curious cats, or children anxious to help. You will need to protect your hands with rubber gloves, and the working surface with newspapers, and be sure that you can measure and handle the ingredients carefully. If you do spill even the smallest amount of lye on your skin, wash the affected area immediately with cold water and dab liberally with lemon juice or vinegar.

In soapmaking the lye is blended with fatty acids that neutralize the alkali. Originally tallow, a saturated animal fat, was used. This could be melted in advance, strained and kept ready in the refrigerator. It is more convenient and more flexible, however, to substitute vegetable oils. Our recipe uses coconut, sunflower and olive oils, but you could use others.

If you do not wish to make soap in this way, there are simpler alternatives. Our ideas include adding your own distinctive fragrances to pure, unperfumed castile soap, and making a "loofah" soap with fine oatmeal.

BASIC SOAP

Make variations of this recipe, as suggested below, and at the end of an absorbing experiment you will have a collection of soaps in a combination of shapes, colors and fragrances.

1¼ cups water
4 tbsp. lye
3 tbsp. coconut oil
3 tbsp. sunflower oil
3 tbsp. olive oil
a few drops of food coloring (optional)
a few drops of fragrant oil or essential oil

Put the water into a glass bowl (do not use a metal container) and very carefully stir in the lye with a wooden spoon. Stir until it dissolves, then set aside to cool – the lye will have heated spontaneously.

Warm the vegetable oils in a small saucepan. When the lye solution and the oils are at the same lukewarm temperature (test by putting your hand underneath the containers), pour the oil slowly and steadily into the lye, stirring constantly. Stir in the coloring if you use it, the flower oils and any other ingredients. Beat the mixture with a whisk, watching carefully for the sudden transition from a shiny liquid to a thick, opaque substance.

Pour the mixture quickly into prepared molds (see below), stand them on a wire cooling rack and leave in a warm, dry place for about 24 hours, until the soap has set. Remove the mixture from the molds, wrap it in waxed paper and leave in a cool place for two to three weeks to harden.

Variations on the recipe

To vary the basic recipe, add three or four drops of an essential oil, bought from a herbalist, or 1½ tsp. of your own fragrant flower or herbal oil. You can also add about 4-5 tbsp. of chopped herb leaves or flower petals.

With so many possible additions, the fragrance permutations are endless. Here are just a few ideas to inspire you. Oil of roses and two or three drops of red food coloring; oil of lemongrass and a pinch of powdered saffron (for yellow soap); homemade elderflower oil and honey for a cream-colored soap; finely chopped marjoram or other herb with a complementary oil and two or three drops (if you wish) of green coloring.

Vary not only the fragrance but the shape of your soap. Use small gelatin and dessert molds, yogurt containers, candle molds or small foil baking pans lined with plastic. You can decorate plain shapes after the soap has set by carving initials, flowers or leaves on the surface. Collect the soap shavings and use them in a metal soap saver.

Colored soaps can be easily made by incorporating food dyes or herbs chopped and mixed with oil.

OATMEAL SOAP

An old country favorite, this variant of the basic soap is not usually given a heavy fragrance.

Basic soap mixture
4 tbsp. fine oatmeal
2 tbsp. clear honey
a few drops fragrant oil or a pinch of sandalwood
  powder, if desired

Make up the basic soap mixture, as described, adding the oatmeal and honey.

ADDING PERFUMES

If you buy unperfumed castile soap, you can transform it by adding any fragrance you please.

8 tbsp. soap flakes
1 tbsp. vegetable oil
1 tbsp. clear honey
a few drops of fragrant oil or essential oil

1. Put the soap flakes into the top of a double boiler or a bowl fitted over a pan of simmering water. Stir with a wooden spoon until the soap has melted.
2. Add the vegetable oil and honey and stir over heat for five minutes.
3. Add the fragrant or essential oil and pour the soap into molds. Leave it to harden, as for basic soap.

LOOFAH SOAP

The notion of lathering yourself with a dry, powdered soap on a loofah is as old as the hills.

3 tbsp. fine oatmeal
3 tbsp. kaolin powder
1 tsp. household borax
a few drops of fragrant oil or essential oil

Stir together the oatmeal and the kaolin powder (fine white clay usually available at a pharmacy). Sift in 1 tsp. household borax and add any fragrant or essential oil of your choice. Pack the mixture into small, lidded pots. These soaps make an unusual gift.

Spicy variations
This type of natural-looking, refreshing soap is best scented with a spicy oil such as patchouli, vetiver or sandalwood.

BEAUTY IN THE BATH

Great beauties of the past have lavished fortunes on ingredients for luxurious baths. Cleopatra made no secret of her fondness for a deep, soothing bath of asses' milk scented with expensive aromatic oils. Catherine the Great sent herb gatherers far and wide throughout Europe and the Far East to discover ever more exotic fragrances for her daily dip. Madame Tallien, an 18th-century French courtesan renowned for her delicate skin, took to bathing in crushed strawberries and raspberries, before rubbing down in rose-scented milk. Madame Pompadour, reputedly one of the greatest beauties of all time and mistress of Louis XV, languished so frequently in baths scented with herbs and strewn with flowers that it worried the king. He feared that this overindulgence would spoil her beauty.

There is some truth in this theory, which generations of schoolboys have been at pains to point out. Some people do find that taking frequent baths has a drying effect on the skin, destroying the acid mantle that forms a protective layer against disease. A simple way to redress this balance is to add a little cider vinegar, sunflower or sesame oil to the water.

Certain herbs and flowers are credited with specific properties, so we can select not just for fragrance, but for the "mood" of the bath we take. Are you in need of complete relaxation? Choose chamomile or lavender flowers. Feeling slightly jaded? Try elderflowers. Just cannot get to sleep? Opt for valerian. Refreshing lovage and mint have been found to be helpful for skin disorders. Marigold leaves have been relied on in times past to soothe varicose veins and fade the surface appearance of capillaries. Rosemary and angelica are considered stimulating, hyssop and houseleek nourish the skin, and both dandelion and nettle leaves can help to cleanse the skin.

TYPES OF OILS

In order to enjoy the fragrance and derive the benefits from aromatic leaves and flowers, you need to extract the oils. There are a number of ways of doing this, apart from the method that involves using a still to produce what is known as essential oil.

First, you can make disposable bags of fresh or dried herbs and flowers – bathtub bouquet garnis – and, for appearance' sake, put each one in a pretty outer covering when you use it.

It is not advisable to scatter tiny blue rosemary flowers, spiky rosemary leaves, minute cream elderflowers, lavender and small petals directly into the bath water, as scraping them from your skin is a very time-consuming, messy business.

Dispersing and floating oils made with petals or herbs and used as "steamers" will add a touch of luxury to bathtime.

A more effective way to extract plant oil is described on page 133, where details are given for a versatile range of fragrant oils. You can use these, or essential oil from herbalists, to make your own fragrant bath oils that have a touch of luxury.

Choose from two types of oil for the basis of the bath oil. If you like one that dissolves in water (and incidentally leaves no ring around the bath) use a dispersing oil. If you prefer to emerge from the bath with a healthy glow, with a glistening preparation still clinging to your skin, choose a floating oil. Almond oil and avocado – rich in skin-nourishing vitamin E – are examples. These oils are usually available at large health food or natural cosmetic stores.

Mix one part of homemade fragrant oil with three parts of the dispersing or floating oil; or sprinkle a few drops of the more concentrated essential oil into the basic type. Mix well, pour the bath oil into attractive bottles and jars, such as those that originally contained herbs and spices, and label them with the names of the fragrances. Carnation and almond, jasmine and avocado, or red rose oil will make a delightful collection for your own bathroom, and would be thoughtful gifts for friends. On each label make a note of the quantity to be used; normally this is 1 tsp. for each bath.

USING VINEGAR

Oil and vinegar may not seem natural ingredients for bathtime beauty, but each has its part to play. The addition of vinegar helps to tone the skin by restoring the natural acid balance and counteracting the alkali effect of soap. To make an infusion, use equal volumes of plant material and white distilled vinegar. Pack the leaves or flowers into a heatproof preserving jar, and pour on an equal quantity of boiling vinegar. Close the jar, leave it on a sunny windowsill or in a warm place and simply shake it once or twice a day.

Strain the vinegar, pressing the plant material against the strainer to extract the maximum fragrance, and pour it into a stoppered bottle. Label the bottle with the fragrance type. Use about ½ pint for each bath. As a skin toner, dilute the fragrant vinegar with eight parts water and use it as an after-bath splash, for all parts of the body.

Herbal and flower infusions in water, known as tisanes, are as beneficial in the bath as they are in the teacup. Make them in the way described on page 162. For maximum relaxation pour one cupful of, say, chamomile tea, into your bath and drink a second cupful while you bathe.

HAIR CARE

Look along the shelves of hair-care products in any shop or supermarket and it is easy to see that those products obviously employing herbs and flowers, and fruits and vegetables, are very popular. Natural ingredients have made a big and welcome return in recent years. It is worth reading the labels carefully, however, for not all the fragrances have their origins in a herb garden. Generally it is only the more expensive products that are as "natural" as those you can make for yourself. So, beyond the therapy and satisfaction, this is another reason for adding hair shampoos, conditioners and rinses to your repertoire.

Herbal and floral shampoos are rewardingly quick and easy to make, and the range of fragrances and benefits is enormous.

BASIC SHAMPOO

This simple soap can be varied by the addition of different flowers: chamomile flowers hold the secret to many a head of beautiful fair hair, as they have a lightening effect; rosemary and sage leaves and lavender flowers – used separately – are time-honored shampoos for dark hair, and marigold petals have just the coppery glow that is needed for red hair.

6 tbsp. castile soap
3 pints water
1 cupful herbs or petals
2 eggs (optional)

1. Grate castile soap. Pour boiling water over herb leaves or flower petals, stir well, cover and set aside to infuse for two hours.
2. Strain the liquid into a saucepan, pressing the plant material against the strainer so that you do not waste a scrap of fragrance.
3. Add the grated soap, whisk over low heat until it has dissolved, cool, then pour into decorative bottles, and label. Shake the bottle well before using. Rinse the hair thoroughly after shampooing. The slightest trace of soap imparts a dull look.

Adding protein
The addition of two eggs, whisked into the shampoo once it has cooled, adds valuable protein. Add only the egg whites for use on greasy hair.

SOAPLESS SHAMPOO

If your hair reacts badly to soap, there is a herb you can use instead, called soapwort. You can buy the dried, powdered root from herbalists.

6 tbsp. powdered soapwort
one cup of herb leaves or flowers, such as elderflowers, wild thyme or lime flowers
3 pints boiling water
one egg (optional)

1. Mix the dry ingredients in a bowl and pour on the boiling water. Stir well, cover and leave to infuse for two hours.
2. Cool and strain. Whisk in an egg for extra protein, if you wish, before bottling and labeling the mixture.

RINSING AND CONDITIONING

After the shampoo comes the conditioner and rinse – and fragrant oils (made according to the instructions on page 133) and fragrant vinegars (page 138) find yet another use. A treatment of warm oil rubbed into the scalp is an excellent conditioner, and any vegetable oil is suitable. If it is fragrant, so much the better. Treated castor oil is good for strengthening the hair, and both avocado and almond oil are rich in protein. And so homemade bath oils (pages 136-38) are just the ticket. For maximum benefit, rub the oil in thoroughly, cover your head with a piece of aluminum foil and then with a towel. Leave for at least 15 minutes, or longer if your schedule allows.

For another after-shampoo conditioner, and one that is easy to make, mix 5 tbsp. plain yogurt with one egg and 1 tsp. fragrant oil or one drop of essential oil. Apply to the hair as described above.

Most people remember being told in childhood that vinegar in the rinsing water leaves hair "squeaky clean." Multiply that benefit by using an infused vinegar (see page 138). As with shampoos, rosemary, sage and lavender vinegars are good for dark hair, chamomile for blond tresses and marigold for brunettes. Comfrey leaves infused in vinegar have disinfectant properties; peppermint leaves, nettle leaves, elderflowers, yarrow flowers (which also help to combat greasiness) and fennel seeds are all recommended to fight or prevent dandruff. Use about 3 tbsp. in the final rinsing water. If you are using dried herbs or flowers, perhaps bought from a herbalist, halve the stated quantities.

COLOR CHANGE

Herbal and floral preparations can be beneficial not only in restoring and maintaining the hair in a healthy condition; they can also be used to alter the color. A paste, made by mixing a plant and water infusion with kaolin or other powder, is likely to be more effective than a rinse when applied to the hair.

To give fair hair that sun-kissed look, try chamomile and rhubarb for the best chance of success. Infuse one cup of chamomile flowers in 1 pint of boiling water (use rainwater if possible since it leaves the hair soft). Stir well, cool, strain and pour the liquid onto 8 tbsp. kaolin powder, stirring constantly. Beat in one egg yolk. Using half the quantity at a time, rub the paste thoroughly into the hair, leaving it on for 20 minutes. Then rinse in clear water, adding a little vinegar to the final wash.

The use of walnut skins for deepening brown hair goes back to Roman times, when infusions of walnut leaves were also used. For the paste method, mix 8 tbsp. chopped walnut skins with 8 tbsp. rosemary and water infusion. Pour the liquid onto 2 tbsp. powdered alum and beat in one egg yolk. Apply half the paste at a time to the hair, using it as it is described above.

Note: When many of these herbal remedies and beauty aids were in widespread use, the herbs and flowers were freely available in gardens and the countryside; this is not now the case, but you can buy many of them in dried form from herbalists.

SKIN TONES

There was an old belief that the finest way to keep skin young and healthy was to bathe it daily in morning dew. And not just any morning dew, but the safe and gentle water gathered drop by drop from the buds and leaves of the hawthorn tree, before dawn on the first of May. Just in case your own supply of this magic moisturizer runs out before next spring, here are some more down-to-earth suggestions for your cosmetics shelf.

There is perhaps only one "secret" of a beautiful skin, and that is to treat it gently, and treat it regularly. An occasional blitz on a tired and undernourished skin can be of minimal benefit, and is quite different from the relaxing, reassuring therapy of a regular routine. If cleansing, toning, nourishing and generally pampering your skin has had a low priority, make a resolution here and now to set aside a few extra minutes night and morning. It is a small enough price to pay for maintaining in good condition the delicate tissue that regulates the body temperature, wards off bacteria, helps the body eliminate harmful toxins and acts as an all-over waterproof.

CLEANSING CREAMS

Cleansing creams are quick and easy to make at home, and can be produced for a fraction of the market price. Dairy produce, purified white beeswax, vegetable oils and honey combined with flowers or floral infusions make some delightful and effective products.

ELDERFLOWER CLEANSING CREAM

In addition to their pleasant scent, elderflowers are particularly soothing to the skin, as are chamomile and lavender flowers and thyme and peppermint leaves.

2 large elderflower heads
1½ cups plain yogurt
3 tbsp. clear honey

These attractive pastel skin creams incorporate saffron, rose and mint for coloring.

1. Put the yogurt and the flowers from the elderflower heads into a small pan and stir over a low heat. Simmer gently for 30 minutes, then cover and set aside for three to four hours.
2. Strain the mixture, then stir in the honey and whisk thoroughly until the mixture thickens. Pour into jars, label and store in the refrigerator.

Elderflower cleansing milk
Make this in a similar way to the cleansing cream, but substitute buttermilk for the yogurt. The preparation does not thicken. Pour it onto absorbent cotton to apply to your face.

ROSE PETAL CLEANSING CREAM

This gives your skin an especially cared-for feeling.

10 tbsp. grated white wax
6 tbsp. sunflower oil
5 tbsp. distilled rosewater
½ tbsp. cider vinegar

1. Melt the wax in a pan, over low heat, and then stir in the sunflower oil.
2. Carefully remove from the heat and stir in rosewater and cider vinegar. Whisk the mixture until it is light and creamy, spoon into jars, and label.

TONERS

Astringents and toners close the pores after deep cleansing and leave the skin with a fresh, tingling feeling.

For oily skins make a simple marigold infusion to use as a toner after cleansing. Pour 2½ cups boiling distilled water onto about 1 cup marigold petals, stir well, cover and leave overnight. Strain, bottle, and label. For an alternative "recipe" for greasy skins, mix 1 tbsp. herbal or floral vinegar and ½ cup rainwater or distilled water.

As a toner for normal and dry skins, mix together ½ cup distilled rosewater, 2½ tbsp. distilled orangeflower water and 2½ cups of glycerin. Bottle and label. Pat the skin with the toner after cleansing.

MOISTURIZERS

The first priority here is to make sure of your skin type: milk steeped in herbs makes a gentle moisturizer for normal or oily skins, but dry skins benefit from heavier, creamy substances.

HERBAL MOISTURIZING MILK

Smooth this onto your face after cleansing and toning, or pour a cupful into a warm bath for a true luxury.

1¼ cups whole milk
3 tbsp. chopped mint, marjoram or parsley
(this last is particularly good for oily skins)

Pour the milk onto the herbs, cover and leave in the refrigerator overnight. Strain, bottle, label, and store in the refrigerator.

COLD CREAM

Use this as a moisturizer night and morning.

½ cup herbal or floral infusion of your choice (if you
use distilled water, it will not be necessary to store
the preparation in the refrigerator)
1 tsp. borax
6 tbsp. purified white beeswax
1¼ cups almond oil
a few drops of essential oil or 1 tbsp. fragrant oil

1. Dissolve the borax in the herbal infusion and set to one side.
2. Melt the purified beeswax and pour in the almond oil, stirring constantly. Stir in the borax solution, remove from the heat, cool and stir until the cream thickens.
3. Add essential or fragrant oil, spoon the cream into jars, and label.

FACE MASK

Spread this mask over face, avoiding the area around your eyes. Soak pads of absorbent cotton in a mild floral infusion to make eye pads (chamomile and elderflower are both good). Rest for 30 minutes, then remove the eye pads, wash off the mask, pat skin dry and continue your usual toning and moisturizing routine.

4 tbsp. finely chopped mint
1 tbsp. clear honey
3 tbsp. milk
2 tbsp. fine oatmeal

Simmer the mint in an equal quantity of water for five minutes. Remove from the heat and stir in the honey, milk and oatmeal. When the mixture has cooled apply evenly to your face.

BURNING BLOOMS

A room scented with perfumed candles, aromatic oil or floral incense takes on an almost mystical quality, as if the whole history of these pungent, all-pervading aromas is re-created in a wisp of fragrant smoke. The art of candlemaking was known to the ancient people of Egypt, China, Japan and Crete. In some civilizations candles bearing marks at regular intervals were used to measure the passage of time. For thousands of years candles have been used to sweeten the air, with aromatic oils and perfume added to the basic ingredients. The "fuel" for the candles varied according to local availability. It may have been oil from seabirds or fish, tallow or animal fat, or plant derived oils, and beeswax (considered the finest).

MAKING CANDLES

The most popular fuel used in candle craft is paraffin wax, a petroleum byproduct. It is mixed with stearin, or stearic acid, a pure form of animal fat that strengthens the wax, makes the candle opaque and slows down the burning rate.

Materials you will need
(for one candle)
9 tbsp. paraffin wax
1 tbsp. stearin
candle dye
1 wick
1 wick rod or toothpick
a small piece of modeling clay
3-4 drops essential oil
or 1 tsp. fragrant oil
or special candle perfume
⅔-cup or ½-pint mold

Melt the wax in a double boiler or a bowl fitted over a pan of simmering water. Cut a piece of the wick 3 inches longer than the mold. You will need thin wick for a candle 1 inch thick, medium wick for a 2-inch candle and the thick type for a candle that is 3 inches

Burning candles sweeten the air with their aromatic oils and perfumes.

or more in diameter. First dip the wick in the melted wax and allow it to dry. Tie it close to the middle of the wick rod or toothpick. Lay the rod or stick across the top of the mold and thread the other end of the wick through a hole in the center of the mold base. Secure it, closing the hole with modeling clay. Stand the mold upright, supporting it in a jar if necessary.

Stir the stearin and as much dye as required into the melted wax, add the fragrance and stir over low heat until well blended. Cool slightly. Pour slowly into the mold and tap it to release any trapped air. As the candle wax sets, depressions will form in the top. Melt any remaining wax mixture and top up the mold as required.

Leave the candle to cool completely. Tap rigid molds or peel back flexible ones to release the candle. Trim the wick and polish the surface of the candle with a soft, dry cloth.

You can buy flexible molds in all shapes and sizes; flower shapes would be appropriate. Improvise by using yogurt containers, mousse and gelatin molds, cups, or even wine-glasses. (You can safely burn a candle inside a glass.)

DECORATING CANDLES

To add a finishing touch to your candles, "paint" them with pressed flowers. Melt some plain wax or the ends of white candles. Place the flowers, petals or leaves against the surface of the candle and affix them with a few quick strokes of a small paintbrush dipped in melted wax. Wild rose petals assembled in a flower shape with a daisy for the center, a ring of fuchsia "bells" around the top of a short, squat candle, or sprays of buttercups twining around a tall, thin candle – make each one unique.

The "candles" burned in Biblical times consisted not of wax set hard around a vertical wick, but of a wick suspended in a dish of olive oil, which is an excellent fuel. Infuse the oil with fresh or dried herbs, flowers or crushed seeds, blended or used individually (full details are on page 133), and burn it in a traditional, boat-shaped burner or a modern glass container, for an aroma that will be as evocative as potpourri.

INCENSE

There are two types of incense. One is a blend of powdered spices, a fixative and flowers; the other is blended with ground charcoal and gum, and shaped into cones or sticks to burn indoors.

POWDERED INCENSE

Sprinkle this incense powder onto glowing charcoal, after first making sure that the room is well ventilated so as to avoid overpowering fumes.

2 tbsp. sandalwood powder
1 tbsp. ground cardamom seeds
1 tbsp. ground cloves
1 tbsp. ground cinnamon
2 tbsp. powdered gum benzoin
2 tbsp. dried lavender flowers (or other highly aromatic flowers)
1-2 drops of essential oil

Mix the sandalwood powder, cardamom seeds, ground cinnamon and cloves together with the fixative, gum benzoin. Stir in the flowers with one or two drops of oil, and store in an airtight tin.

INCENSE CONES

This fragrance is specially appropriate for an Eastern-style party. And for the following morning, or whenever a room needs a refreshing aroma, try an old herbal remedy. Rub dried herbs, such as lavender flowers, rosemary, sage, bay or lemon balm leaves to a powder, and burn them in a fireproof dish. This herbal incense is better than a breath of fresh air.

1 tsp. powdered sandalwood
1 tsp. ground cinnamon
2 tbsp. crushed dried bay leaves
2 tbsp. dried lavender or other aromatic flowers
4-5 drops of essential oil
5 tbsp. gum arabic
1 cup quick-lighting charcoal
2 tbsp. 1 oz powdered gum benzoin

1. Crush charcoal and stir in gum benzoin, sandalwood and cinnamon.
2. Mix gum arabic to a stiff paste with water and stir into the spice mixture. Stir in bay leaves and dried lavender or other aromatic flowers, and four or five drops of essential oil.
3. Stir the mixture well and form it into shapes – cones are traditional.
4. Stand the incense shapes in a warm place for two days to dry, then wrap them in foil to retain the aroma. Unwrap the incense blocks, stand them on a dish, and light the tip to burn.

WRAPPING AND PACKING

When you create your own line of beauty and craft products – fragrant soaps and bath preparations, products to take good care of the hair and skin, or perfumed oils and candles that will waft their aroma into every corner of a room – you build up a selection of pleasing gifts. Perhaps you might like to give elderflower or lime cosmetics to a young girl, a rose-perfumed preparation to a friend with romantic ideals, or spice-perfumed products to an older person.

As soon as you start making beauty products – or even before – start collecting containers for them. For as all the expensive cosmetic houses know to their profit, attractive packaging is all part of the aura. Packaging adds a certain feeling of luxury and both complements and enhances the product.

Try to design a brand image for your line. The bottles and jars of cosmetics and creams will be received with even greater appreciation if they look decorative on a bathroom shelf or dressing table.

Collecting beautiful objects from past eras may be one of your hobbies, in which case turn this to your advantage and seek out old bottles and jars for your cosmetics. Many secondhand junk shops have a section given over to inexpensive but attractive containers dug up in someone's field or garden, often by diligent children hunting for ways to supplement their pocket money. Look out for glass bottles in deep blue, ivy-leaf green and chestnut brown. They would make attractive containers for a line of fragrant hair rinses and oils.

Old soft drink and beer bottles are useful for shampoos and conditioners, and small ink bottles are ideal for oils that have a heavy concentration of herbal or floral aroma.

These lovely old bottles, made of thick glass with a hint of green or brown and with a wealth of texture in the surface, may have lost their stoppers. Except perhaps for the effect this loss has on bargain basement prices, that is not important. Buy a bag of new corks and cut them to make an airtight fit for the bottle necks.

If used containers are not your choice or you do not have the time or opportunity to seek them out, buy plain glass jars or plastic screw-top bottles from

Cosmetics bottles can be "customized" with ribbons, bows, dried and fresh flowers, seedpods and distinctive labels.

houseware shops or drugstores.

However plain or pleasing the container, you should always label all cosmetic and beauty products clearly, with the name and type of the preparation, the quantity to be used and any special instructions. Nothing is more frustrating than to receive a bottle of something that is a delightful shade of pale green and smells divine, but that is completely anonymous, and potentially a hazard.

CUSTOMIZED CONTAINERS

A label gives you another chance to stamp your personality and design preferences on your product. You can buy very attractive printed labels, edged with entwined flowers and leaves, art-nouveau-style tendrils or abstract designs. Write the details clearly in indelible ink – a soapy hand groping for a bottle will obliterate anything written with ordinary inks.

The plainer the bottles or jars, the more you might like to give them a visual facelift. Bows of narrow satin ribbon around the neck of the containers – deep pink- and white-striped ribbon with rose-perfumed products, maybe, and tiny checked gingham with pure white creams – are a simple device, and one that you could easily adopt as your cottage-industry theme.

For a handcrafted look you can wax pressed flowers onto bottles and jars, in the way suggested for decorating candles. Simply melt white wax in a pan, hold pressed flowers, buds or small leaves against the container and brush quickly over it with the hot wax. It will set on impact, holding the decoration in place, and is resistant to cold and warm water splashes.

FLOWERS AS TRIMMINGS

Fresh and dried flowers make the prettiest of present trimmings, adding a final flourish that is uniquely your own. It is easy to wrap a gift package in bright blue or orange paper, tie it around and around with satin ribbon and twine the ends into tight little curls, but gathering up a bouquet, perhaps of small nasturtiums or pansies, to tie onto the box at the very last minute adds a special touch.

Fresh flower trimmings make a few calls on the ingenuity if they are to radiate not only charm but garden freshness too. Bouquets do not present many problems. You can assemble them the day before the event and keep them in water overnight. The next day, if there is to be much of a time lapse between your leaving home and handing over the gift, wrap the stems in damp tissues or absorbent cotton, then bind them tightly with foil to seal in the moisture. Twist a short length of ribbon, that matches the present around the foil if you wish.

Single flowers or a few slender stems can have the luxury of traveling to their destination in a small plastic phial of water. Florists sell them for dress orchids, or you can improvise by using the plastic cap of a felt-tipped pen: fill it with a few drops of water,

Fresh or everlasting flowers teamed with stylish ribbons and careful wrapping will make presents that much more special.

cover the opening tightly with foil pressed firmly all around it, and push in the stems. Bind the stems and the improvised stem holder with ribbon and tie a pretty bow around it. This simple but effective device is a good way of keeping fashion flowers fresh. The pen top is narrow enough to push through a buttonhole or slip under a hatband.

If you are making a present that will be given almost immediately, ribbons of fresh flowers make pretty package trimmings. Tie a box with wide ribbon across the corners, spread the ribbon generously with glue, and press on rows of flat flower heads, touching edge to edge, like a living embroidery. Or dab glue at random intervals over the package, and press on daisies, buttercups, potentilla, celandines, or whatever you have in the garden or vase.

TIMELY PRESENTS

Presents that have a long wait ahead of them – those that will be piled under a Christmas tree or hung on its branches – are best trimmed with lasting flowers. Bouquets of everlasting flowers, helichrysum, statice, gypsophila and honesty make glittering trimmings.

Buy the cheapest brand of colored tape and scatter over it dried flower heads – the perfect way to bring glory to those flowers that have come apart from their stems. Tie bows of ribbon and stick dried flowers in the center and on the trailing ends. Cut out flower shapes or "snowflakes" from white, silver or gold paper doilies, stick dried flowers in the centers – pink and blue hydrangea or larkspur florets or small bachelor's buttons – and scatter them over plain tissue paper wrapping.

Most important of all, keep a floral "button box" for all the snippings and offcuts of dried flowers. A tiny spray of silver honesty, a single rosebud on a tiny stalk, a floret or two of delphinium can all transform a simple package into a beautiful one. It's the thought, and the flowers, that count.

Chapter 5
FOOD AND FLOWERS

Flowers are delicious. That must be the verdict after browsing through old herbal and recipe books, and experimenting with new and long-forgotten recipes.

To regard eating flowers and drinking in their aroma as unusual is a comparatively recent attitude. In times past it was readily recognized that flowers could contribute a combination of spiciness, sweetness, color, shape or aesthetic appeal to dishes as varied as chrysanthemum soup and rose-petal jelly. Indeed, it was only because flowers could earn their keep in the kitchen that precious space was made for them in the garden. Pansies and pinks, mallow and marigolds, primroses and roses, took their rightful place in the country garden alongside other practical crops like peas and parsnips.

THE CULINARY RANGE

But the practice of using flowers for culinary purposes is rooted much further back in time. The Greeks called chamomile, with its little daisylike heads, one of the most versatile flowers, the apple of the earth. The Romans made both rose and violet wines. They also made rose wine without roses, using citron and palm leaves as alternative flavorings. In the Middle Ages, hippocras, a highly spiced wine – renowned for its supposed aphrodisiac properties – was scented with clove carnations, the sweet-smelling border pinks.

In those days it was almost a case of "marigolds with everything." Mutton broth, "ordinary pottage," stewed beef, crab salad, marigold bread, buns and sandwiches, baked and boiled pudding, pickled buds, crystallized petals, conserve, cordial, wine and an after-dinner digestif, all benefited from the flavor and appearance of these spicy flowers.

Exploring the culinary opportunities that exist in the flower garden (or in a well-stocked nursery) is as exciting as opening an imaginative and colorful cook book for the first time. The effects you can create by using flowers as decorations and garnishes are there for all to see. But what of their flavor; how do you judge that? The answer can only be found by testing and tasting.

Pick off a petal here, nibble the corner of a leaf there, and assess the flavor. It must be stressed that you should never taste any part of any plant unless you know it is safe to eat (there is a more detailed note on page 150).

For new enthusiasts to the art of flower tasting, there are some delightful surprises in store. Chrysanthemum, nasturtium, gladiolus, lilies and marigolds are among the spiciest of blooms. In China, white chrysanthemums have been cultivated for culinary purposes for over 2000 years. The petals are used with pickling spices to preserve fish, tossed in stir-fries of beef, chicken and vegetables, floated on rice wine, and sprinkled on cooked dishes as an elegant, waxlike garnish.

Both nasturtium and gladiolus have shape as well as piquancy in their favor. The trumpet-shaped flowers may be dipped in batter and deep-fried, or used as eye-catching natural containers, filled with salads, creams and sauces. The flowers may be infused in oil and vinegar to give a hot pepperiness to sauces and salad dressings, or sliced and tossed, raw, in salads. Both flowers may be added, just for flavor, to soups and stews (they emerge from simmering liquids limp and bedraggled and should then be discarded). In addition, young nasturtium leaves give a lift to green salads and the buds may be pickled and used in a similar way to capers.

In Mexico the lily, or shell flower, is the star of the kitchen. The flowers are used in great abundance to pep up sauces. In China dried or pickled lily buds – picked when they are no more than 2 inches long – add spice to dishes such as braised duck (with which they have a special affinity), steamed whole fish and fried rice. And in Norway lily flowers are steeped in brandy and sugar to make a liqueur.

On a still summer evening, cool drinks made with fragrant flowers will be a welcome refreshment.

Roses take pride of place among the sweet-smelling, sweet-tasting flowers, and have so much to offer that it seems a crime to let their petals fade and flutter, unused, to the ground. Other flowers that have a taste for confectionery, and can be used in many of the same ways, include clove carnations, broom, honeysuckle, jasmine, lavender, mallow, pansy, primrose, acacia, violet, sweet cicely, meadowsweet and elderflower.

INFUSING IDEAS

Draw out the flavor by infusing the flowers or petals in honey for toppings, baking and party spreads. Dry the flowers in sugar to carry the sweet, subtle aroma through cakes, cookies and desserts. Steep them in milk or cream to give a hint of luxury to pancakes, waffles, ice cream, fruit mousses, crême brulée, baked custard and rice molds. Infuse the fragrant petals in syrup to make delectable sorbets, fruit salad dressings and drinks, or give them a place of their own in the kitchen cupboard and make a range of fragrant jams and jellies. Our recipes for rosepetal jam, honeysuckle and peach cheese and others are intended as just the tip of the inspiration iceberg.

Flower fritters for dessert were considered a great delicacy at the beginning of this century. A wisp of light batter aerated with beaten egg white enclosed clusters of acacia, sweet cicely, meadowsweet and elderflower, or single mallow and pansy flowers. Present-day equipment for deep-frying, with temperature control and automatic timing, helps you to make these desserts in moments.

GARLANDS OF GARNISH

Of the many ways to prepare flowers for the table, perhaps candying is the most romantic and evocative. But there are other ways. Flowers make the most eye-catching of instant garnishes and decorations; even a couple of blooms and leaves plucked from a flower arrangement or an indoor plant will transform an everyday dish.

Lining a serving plate with leaves is a good start. Shiny variegated ivy leaves make an attractive pattern beneath a selection of cheeses, and yellow marigolds scattered on bunches of black grapes between the wedges and rounds intensify the dairy coloring – and serve as a reminder that the petals are especially good in egg and cheese dishes of all kinds.

Fresh fruit looks good with leaves. Line a plate or a pedestal cake stand with leaves – vine and hop foliage is perfect – and group polished apples, downy peaches, clusters of black cherries, fresh dates, halved kiwi fruits and tiny bunches of grapes. For a finishing touch add three or four crimson roses, echoing the color of the grapes, or a spectacular head of deep red geranium.

Choose scented geranium leaves to line a dish of ice cream or decorate a mousse, and contrast or complement the color of the food with flowers. A blend of pale green and scarlet tobacco flowers with pistachio cream, a stem of alstroemeria decorating an apple mousse, a spray of honeysuckle with apricot sherbet, work visual wonders.

Chopped fresh herbs and sprays of herb leaves can seem a little uninspired for garnishing vegetables and savory dishes. If home-grown green beans are a proud feature of the menu, give the dish a dazzle with a spray of the scarlet flowers. After all, at one time, before people realized how delicious the pods were, the plants were grown for the joy of their showy flowers. Garnish a dish of green peas or snow peas with a sense of humor, and a spray of sweet peas. The flowers have a familiar flavor and there is no mistaking the family resemblance.

Scattering marigold and chrysanthemum petals on casseroles for a touch of spice certainly proves that flowers and savories go well together. Garnish a dish of roast turkey with a single vibrant head of red geranium, a visual match for the cranberry sauce or stuffing; a chicken with flowering thyme, marjoram or rosemary to complement the filling; or a plate of cold roast beef with a deep red poppy or a single dahlia.

In Mexico there is a saying that the lily provides two feasts, one for the eye and the other to allay hunger. The same might be said of many flowers.

A word of warning

It is unwise to eat any part of any plant unless you know it is safe to do so. Some plants are poisonous. Most poisonous are the purple berries of deadly nightshade (*Atropa belladonna*). The fruits of woody nightshade (*Solanum dulcamara*), bryony, dogweed, spindle and yew are harmful. All parts of thornapple, oleander and henbane plants are poisonous. So, too, are the leaves of tomatoes, rhubarb and potatoes.

Do not be tempted by a flower's good looks. For example, foxgloves (*Digitalis purpurea*), wild and cultivated aconites, pink-flowered valerian and wormwood are all poisonous. Beware, too, of look-alikes. Several plants that look like Queen Anne's lace are harmful. These include hemlock, cowbane, fool's parsley and other varieties.

CAKE DECORATIONS

Real or model edible flowers make attractive decorations for cakes and desserts, and we explain both how to candy real petals and how to make decorative flowers from almond or fondant paste.

CANDIED FLOWERS

Candied (or crystallized) flowers can be made in two ways. You can paint the petals with egg white and frost them with sugar, but this method ensures the flowers will keep for only a couple of days. For longer-term keeping you need to seal the petals in a solution of gum arabic crystals. Then, stored in an airtight container, they will keep for up to one year.

Small flowers or separate petals are most effective. Choose from violets, chamomile, primroses and other primula, small pansies, sweet peas, border pinks, cowslips, borage and fruit blossom, sprays of broom, heather, hyacinth and lily-of-the-valley, and separate rose and carnation petals.

Do not forget leaves as you build up a sugar-coated collection. They flatter the flowers considerably. Small sprays of young mint and lemon balm are among the most effective.

For the long-keeping method, put 1 tbsp. gum arabic crystals into a jar with 3 tbsp. orangeflower water. Cover the jar, shake it vigorously and set aside for three days where you can see it so that you remember to shake the jar frequently. It is most important that the crystals dissolve completely and form a syrupy consistency.

Wash and thoroughly dry the flowers and leaves. If they are at all damp, the solution will not adhere to them. Using a very fine camel hair paint brush, paint every part of the flower or petal with the solution. Hold the flower with tweezers, sprinkle it lightly with fine granulated sugar and shake off any excess.

Place the frosted flowers, petals and leaves on aluminum foil and dry them in a warm place, for about 24 hours, or until they have lost their suppleness and feel really firm.

Transfer the candied flowers onto sheets of foil or waxed paper in an airtight tin. Store them carefully where they will not be disturbed. They are very brittle, and break easily.

FONDANT PASTE

This is the sugar paste from which you can make realistic flowers for cake decorations.

1 cup sugar
⅓ cup water
1 tbsp. liquid glucose, or ¼ tsp. cream of tartar dissolved in 1 tsp. water
flavoring (see note, below)
2-3 drops red or yellow edible food coloring (optional)
up to ⅔ cup confectioners' sugar, sifted

1. Put the sugar and water into a large, heavy-based pan and dissolve the sugar over low heat. Bring the syrup to a boil and add the liquid glucose or cream of tartar solution. Boil the syrup for 10-15 minutes, or until it reaches 240°F.
2. The temperature of the sugar is important to the final consistency of the paste, which must be opaque and firm enough for modeling. If you do not have a sugar thermometer, test the syrup in this way. Drop a little into a cup of cold water. It should form a soft ball on contact with the water, and flatten when it is removed.
3. When this temperature is reached, remove the pan from the heat, wait for the bubbles to subside, then quickly beat in the flavoring and coloring. Pour the syrup slowly into a heatproof bowl. Leave it to rest for one minute and, when a skin has formed on top, stir it vigorously with a wooden spoon until it becomes opaque and firm. Dust your hands with confectioners' sugar and knead the paste until it is smooth and free from cracks.
4. Place the fondant in a small bowl, cover with a damp cloth and set aside for one hour.
5. If you wish to make the paste in advance of modeling it into flowers, wrap it closely with waxed paper and store it in an airtight tin. It should keep for up to one year.
6. Divide the fondant into several pieces and knead each one. The warmth of your hands will make the paste malleable. Sift onto each piece just enough confectioners' sugar to make the paste stiff enough for modeling. You will readily be able to judge when it has the right "feel."

Flavoring
To give molded rose petals the appropriate fragrance, add two or three drops oil of rose petals or 1 tsp. distilled rosewater.

ALMOND PASTE

New enthusiasts to the craft of modeling edible flowers might find almond paste easier to work than fondant paste.

¼ lb. ground almonds
¼ cup granulated caster sugar
2 cups confectioners' sugar, sifted, plus extra for
   dusting
¼ tsp. almond extract
½ tsp. lemon juice
1 small egg, lightly beaten
2-3 drops red or yellow edible food coloring
(optional)

1. Mix together the ground almonds and the sugars. Add the almond extract, lemon juice and egg, and mix to form a paste. Add food coloring, but if you are making green leaves, set aside some of the mixture before this.
2. Dust your hands with confectioners' sugar and knead the paste until it is smooth and pliable. If you do not wish to mold the paste right away, close-wrap it in foil and store it in the refrigerator.
3. When you are ready to mold the paste, sift a very little of the extra confectioners' sugar onto it and knead until it is smooth.

MOLDED ROSES

Decorating a cake for a special party – perhaps a birthday or anniversary – with a bouquet of realistic molded roses must be one of the most charming ways to "say it with flowers." You will soon get the knack of shaping petals and assembling them into roses. Add a few drops of oil of roses to the mixture, and you can even enjoy the scent of the flowers too.

almond paste or fondant paste
confectioners' sugar (to dust)
one beaten egg white (optional, for use with fondant
paste)

1. Roll the almond paste or fondant paste into a long roll about ½ inch in diameter and cut it into ½-inch slices. Keep the paste covered with foil while you shape each petal, to prevent it from drying out. Lightly dust a piece of waxed paper or foil with confectioners' sugar and fold it in half.
2. Roll one slice of the paste into a ball, place it between the paper or foil and "brush" it gently with the back of a spoon to make a petal shape. Press the paste so that the top of the petal, the widest part, is farthest away from you. The paste should be very thin at this top edge, and gradually thicken toward the petal base.
3. Peel the petal from the paper or foil, and make more petals in the same way, dusting the paper or foil with more confectioners' sugar. Petals made from almond paste may be shaped right away; fondant petals should be left to rest for 10 minutes before shaping.
4. Using your fingers, gently roll the first petal to form the center of the flower, then press on five or seven petals, each one slightly overlapping, to build up the flower shape. Trim the flower base neatly.
5. If fondant petals do not hold their shape well, work a little more sifted confectioners' sugar into the mixture. If the petals do not readily stick at the base of the flower, dab on a little lightly beaten egg white.
6. Make more roses in the same way. Leave fondant flowers in a dry place for one hour to set before arranging them.
7. Arrange the flowers in a cluster in the center of an iced cake. Stick them in place with a dab of fondant paste.
Note: To make leaves as well as flowers from the almond paste or fondant paste, separate a little and color it with green edible food coloring. Roll the paste to about ¼ inch thick and mark the veins with a knife point.

CHOCOLATE ROSE LEAVES

Cover rose leaves with a thin layer of chocolate, and you transfer all the beautiful and intricate natural markings to attractive and edible decorations. Arrange the chocolate leaves in sprays with candied or molded flowers to decorate cakes and desserts, or serve them with coffee, as part of a selection of petits fours.

2 oz. dark or semisweet chocolate
12-15 well-marked rose leaves

1. Melt the chocolate in the top of a double boiler or in a bowl fitted over a pan of simmering water.
2. Wash and dry rose leaves. Do not choose very young ones, as they tend to wilt under the strain.
3. Using a small paint brush, coat the underside of each leaf with the melted chocolate. Place them carefully, coated side up, on a baking sheet and leave them in a cool place for about two hours to harden. Store them in an airtight tin and, just before you wish to use them, peel off the rose leaves.

SUGAR BOUQUET

Flowers made from gum paste will make pretty cake decorations. They are modeled in a way similar to that used for kiln-dried porcelain, except that these confectionary blossoms set to a very hard finish without heat. They are especially suitable for wedding, anniversary and other celebration cakes because they can be stored almost indefinitely – a lasting keepsake of a memorable occasion. They cannot be washed, however, so it is best to keep them under glass, where they will be free from dust. A bouquet of these delightful gum paste flowers placed under a Victorian-style glass dome makes a charmingly romantic arrangement in the home.

MODELING PASTE

2 tsp. powdered gelatin
3 tsp. hot water
2 tsp. cold water
1 lb. sifted confectioners' sugar
3 tsp. tragacanth
3 tsp. liquid glucose (melted)
3 tsp. cooking fat (melted)
1 egg white

Making the paste

Pour the confectioners' sugar into a large bowl and mix with the tragacanth; warm in oven at 200°F for approximately 30 minutes.

Put warm and cold water into a small basin and sprinkle in the gelatin; stand in a saucepan of hot water and stir until dissolved. Add the melted glucose and melted cooking fat. Add this to the warmed sugar and tragacanth, and beat well in a mixer until the paste is very white and pliable. When cold, seal in a plastic bag and then in an airtight container, and put in the refrigerator for at least twelve hours. Leave at room temperature for at least two hours before using. The paste will feel quite hard but will soon become pliable in the warmth of the hand.

Note: This modeling paste can be bought in a powder form to which you add water only; it must be thoroughly mixed until white, smooth and pliable.

Delicate swags and clusters of flowers in peach, apricot and white complete the cake decoration.

Materials you will need for the decoration

modeling paste in white, apricot and green
yellow stamens for lilies
2 pieces of foam, one piece very thin, the other about
    1 inch thick
block of florist's plastic foam to support flowers
    while drying
florist's tape (can be used to cover the stems but for
    a wedding cake white or silver looks more delicate)
leaf mould
wire in various thicknesses (medium and heavier
    gauges) and very thin silver wire
small rolling pin
small dowel, pointed one end, rounded the other
large and medium rose petal cutters
large rose leaf cutter
stephanotis cutter for the small lilies
calyx cutter

(Large flowers can be dried on a wire coathanger; bend the wire at base of stem to form a hook and hang the flower on the crossbar upside down.)

THE ROSE

Materials you will need
**colored paste in three shades of apricot; also a small
piece of green paste
heavy-gauge wire
large and small rose cutter
calyx cutter**

Take a small piece of modeling paste and knead until
pliable. Add apricot coloring until well blended; halve
this and mix with the same quantity of white. When
well blended, halve this, and mix with a piece of white
the same size. Mix well. Always keep the paste that
is not being used under a glass tumbler to prevent it
from drying. There are now three shades of apricot.

Take a piece of wire and make a hook at one end.
Make a fat teardrop shape using a piece of the deepest
colored paste to form the foundation of the rose; dip
the hook into egg white and insert wire through the
apex of teardrop and pull through, embedding the
hook in the base.

Layer 1. Using the same colored paste, roll out
very thinly and cut a petal with the small petal cutter,
then smooth and stretch slightly with fingertips. Paint
half of the petal with egg white and wrap it around
the teardrop bud, positioning the apex just below top
of petal. Gently furl the edge of petal with fingertip
and leave to dry completely.

Layer 2. Cut another two petals in the same way
and smooth with fingers, making the edges of the
petal very thin. Moisten the base of each one and
attach to the bud, wrapping it tightly and curving
under the edges of each petal.

Layer 3. Cut three petals in medium-toned paste;
roll out very thinly and shape in the palm of hand;
roll the edges with the thin stick, which will give a
ruffled appearance. Attach to the flower head with
egg white, overlapping each petal.

Layer 4. Repeat layer 3, but lay a piece of cotton
between the petals so that they stand away from the
last layer.

*Give maximum height to the wedding cake by supporting
the top tier on 6 smaller columns.*

Layer 5. With the palest shade of paste and large
petal shaper, cut five petals very thinly. Roll the edges
with the small stick until they are ruffled. Attach each
petal with egg white to the base of flower head, not
overlapping. Curl the edges of each petal and again
place cotton between layers. When the rose is hung
upside down to dry, the leaves will not collapse. Look
at a real rose and copy the way the petals are formed.

154

CALYX

Take a piece of green paste and roll out very thinly; imprint with the calyx cutter and lay on the thin piece of foam. Gently stretch each sepal with the round end of the stick and then brush center with egg white; pierce the centre with rose stem and slide up to the base of rose; gently turn back the sepals to give a natural appearance. Roll a small ball about the size of a small pea, and push the stem through the center, sliding this up behind the calyx.

THE LILY

stephanotis cutter
yellow stamens
wire medium gauge
calyx cutter
white and green paste

Roll out white paste very thinly and cut flower with stephanotis cutter. Lay it on the thin foam and stretch each petal with the rounded end of thin stick. Dampen edge with egg white and roll up evenly to form a trumpet. Make a hook at one end of a piece of wire, and bend 4 stamens in the center; attach to hook and push wire through the center of trumpet, embedding the hook in the base of lily.

Make a calyx as for the rose, but after stretching sepals, lay it on the thick foam and press the rounded end of the stick down into the center to form a cup.

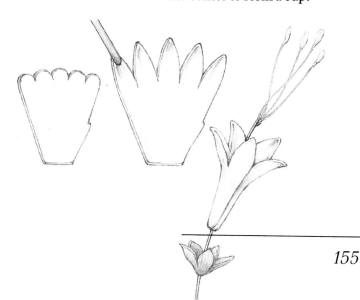

BABY'S BREATH

white paste
very fine silver wire

Take a small piece of white paste about the size of a lentil. Roll to form a ball and flatten with a thin stick. Roll it very thinly to form a rough circle about the size of a small shirt button. Make a hook in the end of a piece of fine wire and gently pull through the center of circle; moisten with egg white and fold in half. Moisten again and fold the right corner to the center, and the left corner over the top – to form a small bud. Stick this into the block of plastic foam. Gather 3 of these little buds, and bind them with a piece of wire about 1½ inches from each bud and bend them outward.

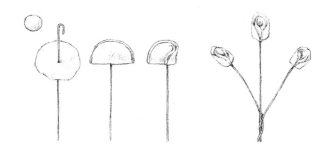

LEAVES

rose leaf
leaf mold
wire

Roll a piece of green paste very thinly and cut a leaf with the rose leaf cutter. Lay it on a leaf mold and lightly roll it on both sides to imprint the veins and midrib. Moisten base of leaf with egg white and lay a piece of medium-gauge wire on the leaf and pinch together. Bend the leaf and twist the point so that it looks natural. Push the leaf into the plastic foam and leave it to dry.

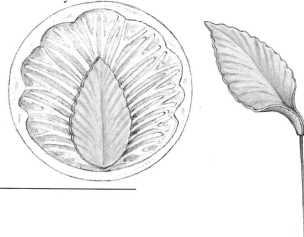

A CHOICE OF DESSERTS

Pansies and border pinks, chrysanthemums and carnations, deadnettles and daisies and many other garden flowers were once regularly used in cooking. In his 16th-century herbal John Gerard wrote that the tiny flowers of red deadnettle were "baked with sugar, such as roses are"; primroses were combined with flour rice, almonds, honey and saffron to make a golden pottage; meadowsweet flowers were used as a honey substitute; pansies decorated jellies and custards; sky-blue borage flowers were baked in a custard tart, and cowslips were included in baked cream. We hope these recipes will help revive the custom.

ROSE PETAL SOUFFLE

A light dessert with a subtle but unmistakable aroma.

SERVES 4
1 tbsp. powdered gelatine
3 tbsp. warm water
3 eggs, separated
½ cup rosepetal sugar (see page 157)
1 tbsp. distilled rosewater
2-3 drops red edible food coloring (optional)
1¼ cup heavy cream, whipped
TO DECORATE
10-12 fragrant pink rose petals, washed and dried
1 egg white, lightly beaten
1 tbsp. granulated sugar

1. Dissolve gelatin in water, stirring frequently.
2. Whisk the egg yolks and sugar until thick and creamy. Pour in the dissolved gelatin, beating constantly. Stir in the rosewater and coloring, then the whipped cream. Leave mixture in a cool place.
3. When the mixture begins to thicken, beat the egg whites until they are stiff. Fold them into the mixture, using a metal spoon. Turn the dessert into a serving dish, cover and chill for two hours.
4. To make the decoration, dip the rose petals in egg white and shake off the excess. Dip them in sugar to frost them. Place them on a piece of foil and leave in a dry place for at least two hours to dry. Arrange the petals on the soufflé.

PANSY GELATIN

In this turn-of-the-century recipe pansy flowers are set in gelatin.

SERVES 4
2 tbsp. powdered gelatin
3 tbsp. warm water
½ cup granulated sugar
thinly pared rind of 2 large lemons
⅔ cup lemon juice
½ stick cinnamon
1¾ cups water
whites and crushed shells of 3 eggs
6 tbsp. sweet sherry
4 pansy flowers, washed and dried
ladies' fingers cookies, to serve

1. Dissolve the gelatin in the water, stirring it frequently.

2. Put the sugar, lemon rind, lemon juice and cinnamon stick into an enamel or stainless steel saucepan, pour in the water and stir over low heat to dissolve the sugar. Remove the cinnamon stick and stir in the dissolved gelatin.

3. Beat the egg whites and shells together until they are frothy, pour them into the pan, add the sherry and whisk over the heat until the mixture boils. Remove from the heat and leave until the froth sinks.

4. Bring the mixture back to a boil and leave it to cool twice more. Line a strainer with two layers of cheesecloth and strain the gelatin. Divide the gelatin between four serving glasses, reserving about 8 tbsp. Chill to set the gelatin, place the pansies face upward and spoon the reserved gelatin over.

MARIGOLD AND APPLE TART

Bright as the summer sunshine, and absolutely delicious, this open apple tart is adapted from a recipe of the 1920s.

SERVES 6.
2¼ lb. green apples
½ lb. ready-made pie crust pastry
juice of 1 lemon
3 large eggs
4 oz. marigold-flavored granulated sugar
  (see below)
⅔ cup heavy cream
petals of 2 orange marigolds, washed and patted dry
3 marigold flowers and 3 small marigold leaves, to
  decorate

1. Peel, core and quarter the apples and drop them at once into a bowl of water containing lemon juice. Heat the oven to 400°F. Grease a 9-inch pie pan and a cookie sheet.

2. Roll out the pastry and use it to line the pie pan. Trim the edges and prick the base.

3. Drain and dry the apples and make slits close together on the rounded side. Arrange the apples, rounded sides up, in the pie pan. Bake, uncovered, in the oven for 20 minutes.

4. Beat the eggs and sugar until pale and frothy. Stir in the cream and marigold petals.

5. Remove the tart from the oven, pour in the custard and bake for another 20-25 minutes, until the custard is set.

6. Leave the tart to cool, transfer it to a serving plate and decorate with the marigold flowers and leaves.

Marigold-flavored sugar

Mix washed and dried petals with twice their volume of sugar, spread the mixture on cookie sheets and place in the oven at the lowest setting for one and a half to two hours, turning frequently, until the sugar has absorbed the moisture from the petals and dried. Cool and store, sifting the sugar before using. (This recipe and method can be used for any flower-flavored sugar.)

A Victorian recipe for a glistening gelatin decorated with pansy heads makes a special treat.

FLOWERS IN THE PANTRY

For generations there have been few tasks more satisfying than stocking up the pantry shelves with jams, jellies, chutneys, cheeses and butters made from the fruits of the orchard. It is only in recent times that a line has been drawn between produce that is edible and that which is supposed to be purely decorative. In these recipes – for rose petal jam, honeysuckle and peach cheese, pickled gillyflowers, and others – we have fudged this line; if a flower is edible and has a pleasant fragrance, experiment with ways to capture it now, and enjoy it later.

ELDERFLOWER AND WHITE CURRANT JELLY

Elderflowers and white currant together well. Wrap a few of the flower heads in cheesecloth when you make white currant jam, bake or poach the fruit, and liquefy it.

MAKES ABOUT 3¼ LB.
4½ lb. white currants
4 cups water
4-6 large heads elderflower
granulated sugar (see method)

1. Put the white currants, water and elderflowers into a large pan, bring to a boil and simmer gently until the fruit is cooked. Mash the fruit occasionally with a wooden spoon.
2. Strain it for several hours through a cheesecloth hanging over a bowl. Do not squeeze the cloth to hasten the dripping process, as that will cloud the jelly.
3. Measure the fruit juice and weigh 2 cups sugar to each 2½ cups juice.
4. Put the fruit juice and sugar into a large pan, heat gently until the sugar has dissolved. Then bring to a boil, stirring occasionally. Boil rapidly for 15 minutes, or until the jelly reaches setting point.
5. Pour the jelly into warm jars, cover, cool and label. Store in a cool, dry, dark place.

Honey will be enhanced with an infusion of flowers or herbs.

ROSE PETAL JAM

Use this preserve sparingly, as a spread on toast, a filling for cakes or spooned over ice cream.

MAKES ABOUT 2¾ LB JAM.
1 lb. fragrant rose petals
6-8 rose-scented geranium leaves
3 cups (1½ lb.) granulated sugar
1¼ cups water
3 tbsp. lemon juice

1. Pick off the white bases from the rose petals. Discard any damaged or discolored ones. Wash and dry the petals and geranium leaves and place them in a bowl. Pour in half the sugar, stir well, cover and leave for two days, stirring occasionally. Discard the geranium leaves.
2. Put the remaining sugar into a pan with the water and lemon juice and heat gently until the sugar has dissolved.
3. Add the rose petal mixture, stir well and heat gently. When the sugar has dissolved, bring to a boil and boil rapidly for 20 minutes, stirring occasionally, until the jam reaches setting point.
4. Pour the jam into warm jars, cover, cool and label. Store in a cool, dry, dark place.

FLOWER HONEY

You can infuse inexpensive blended honey with the flowers of your choice and create a unique spread for toast.

1 lb honey
1 cup fragrant petals or leaves
water to boil

1. Divide the honey between two jars. Wash and dry petals such as rose, violet, honeysuckle, lavender, jasmine, broom, carnation and lime flowers, or scented geranium or other fragrant leaves, and divide them between the jars. Stir well, cover the jars and stand them on a trivet or folded cloth in a pan.
2. Pour in cold water up to the neck of the jars, bring to a boil and simmer for 30 minutes. Remove the jars from the pan, leave to cool and set aside for 24 hours.
3. Heat the jars again and strain off the petals. Pour the honey into small, clean jars.

   This extra-special honey is good for baking, stirred into plain yogurt, spooned over ice cream or spread on toast.

FLOWER SYRUP

Fragrant syrup is a delicious standby. You can use it as a dressing for fruit salad, for poaching fruit, in baked puddings, to make sherbet and in wine cups.

2 cups granulated sugar
1¼ cups water
1 cup fragrant petals or leaves

1. Place the sugar in a pan with the water. Add fragrant petals or leaves – as suggested for flower honey – and stir well. Heat gently until the sugar has dissolved, bring to a boil and simmer for 10 minutes. Cool the syrup, then strain and bottle.
2. Store the syrup for up to two weeks in a refrigerator. For longer keeping, sterilize the bottled syrup by bringing it to a boil in a pan of water and boiling for 10 minutes. Cool, label and store in a cool, dark place. Or freeze the syrup in a container with a lid.

HONEYSUCKLE AND PEACH CHEESE

The term "cheese" denotes the texture of this preserve, which is set in decorative molds and cut in slices to accompany cold meats, poultry and game, or to serve after dinner, with coffee. Honeysuckle and peach cheese is specially good with ham.

MAKES ABOUT 2 LB. PRESERVE.
2 lb. peaches
1¼ cups water
1 cup honeysuckle flowers, washed and drained
juice of 1 lemon
granulated sugar (see method)

1. Halve and slice the peaches, but do not peel them. Crack six pits and discard the outer shells. Put the peaches and the kernels in a pan with the water, honeysuckle flowers and lemon juice. Bring to a boil and simmer for 30 minutes, until the fruit is tender.
2. Press the fruit through a nylon strainer and weigh the pulp. Add an equal weight of sugar and cook over low heat, stirring frequently, until the paste is thick and dry, and a spoon drawn through it leaves a trail. This may take 45-60 minutes.
3. Spoon the preserve into lightly greased decorative molds such as gelatin and mousse molds. Cover the surface closely with waxed paper, then cover the molds with transparent "jam pot" covers. Label and store in a cool, dry, dark place.

"ORANGE FLOWER" BUTTER

Highly perfumed flowers such as orange blossoms and Philadelphus make a delightful spread that has the consistency of soft butter.

MAKES ABOUT 2 LB.
2 lb. green apples, chopped
1¼ cups water
1 cup orange or Philadelphus blossom, washed and drained
granulated sugar (see method)
2 tbsp. orangeflower water

1. Put the apples, water and blossoms into a pan, bring to a boil and simmer for about 30 minutes, until the apples are soft. Press them against the sides of the pan occasionally.
2. Press through a nylon strainer and weigh the pulp.
3. Weigh 1½ cups sugar to each 1 lb. pulp. Put the sugar, fruit purée and orangeflower water into a pan and heat slowly until the sugar dissolves. Simmer over low heat, stirring frequently, until the preserve is like thickly whipped cream.
4. Spoon it into warmed jars, cover, label and store in a cool, dry, dark place.

PICKLED GILLYFLOWERS

This recipe, inspired by an 18th-century notebook, gives a method for preserving clove carnations. You can pickle rosebuds in a similar way, substituting thinly pared lemon rind for the cinnamon.

FILLS FOUR 4 OZ. JARS
1 cup clove carnations
1½ cups white wine vinegar
½ cup sugar
½ stick cinnamon
2 blades mace

1. Cut off the calyx and the white base of the petals. Thoroughly wash the flowers and toss them on absorbent paper towels to dry.
2. Put the sugar, vinegar, cinnamon and mace into a pan, heat slowly until the sugar dissolves, then bring to a boil, stirring occasionally. Boil for three minutes, then cool.
3. Pack the flowers into small, sterilized jars and pour over the vinegar to cover them. Close the jars with lids (don't use metal lids), label and store in a cool, dry, dark place.

AROMATIC DRINKS

There is more to sipping a cup of herbal or flower tea than meets the palate. Tisanes were well authenticated by herbalists two and three centuries ago, and were the main form of medicine made at home.

Chamomile tea was considered a gentle and soothing sedative; lime flower tea was recommended for its effect on the nervous and digestive systems; elderflower, yarrow and lungwort tisanes were all thought to relieve bronchitis and colds; marigold tea was taken in the morning to relieve melancholy and lavender flower tea was said to be a sure remedy for nervous headaches – some claims were made for its properties as an aphrodisiac, too.

Cool, refreshing drinks, made with flowers from the garden, are a summertime bonus.

TISANES

With so many credentials to their name, it is no wonder that herbal and flower teas are popular again. You do not necessarily need to search for materials in the garden. First your herb and spice rack may offer possibilities. Dried mint, thyme, rosemary and marjoram leaves, and fennel and fenugreek seeds all make tasty teas. Buy other dried leaves, flowers and seeds from herbalists or health food shops. For convenience sake use tea bags if they are available.

If you use a teapot, keep one specially for tisanes; otherwise the tannin in "ordinary" tea might spoil the delicate flavors. For each cup of tisane allow about 2 teaspoons lightly bruised fresh leaves or flowers, and 1 teaspoon if the materials are dried. Trial and tasting will tell you whether you prefer to use more or fewer leaves in the future.

Pour in 1 cup boiling water, stir well and cover the pot to prevent too much of the volatile oil from being lost in evaporation. Stand the pot in a warm place, leave it to infuse for 8-10 minutes, and then strain it. Tisanes may be sweetened with honey and decorated with a thin slice of lemon, orange or lime.

LINDEN TEA

Surely home-based medicine at its most delicious! If you do not have the chance to gather the sweet-smelling lime flowers in midsummer, the tea will be just as good if you make it with dried flowers. The tisane is calming, relaxing and helps to induce sleep.

MAKES ONE CUP
2 tsp. fresh or 1 tsp. dried lime flowers, lightly crushed.
1 cup boiling water
honey, to sweeten (optional)
1 thin slice lemon

1. Put the flowers into a pot reserved for the tisanes, pour in the boiling water, stir well and cover the pot. Set it aside in a warm place for 8-10 minutes.
2. Strain the tea and taste it. Sweeten it with honey if you wish, and decorate it with a slice of lemon.

CLOVE CARNATION TEA

If ever you have the chance, look at a catalog of specialty teas and gather ideas for your own blends, adding fragrant flowers and leaves to scent leaf tea. Rose petals, red clover flowers, nasturtium flowers, lemon-scented geranium leaves all make interesting teas with a difference. For this sample recipe, which makes three cups, our choice was clove carnation, or clove-scented pink, a highly perfumed flower that in Elizabethan times was thought to have the power "to comfort the heart."

MAKES 3 CUPS
3-4 tsp. Earl Grey tea
a handful of clove carnations
2½ cups boiling water
sugar or honey, to sweeten (optional)
3 thin slices orange

1. Put the tea and carnations into a pot, pour in the boiling water, stir well and cover the pot. Leave it in a warm place for 10 minutes.
2. Sweeten the tea with sugar or honey to taste, and serve with a slice of orange.

ELDERFLOWER "CHAMPAGNE"

Serve this delicately flavored, nonalcoholic drink well chilled and with a slice of lemon.

MAKES 10 PINTS
6 large heads elderflower
3 cups granulated sugar
2 tbsp. white wine vinegar
10 pints water
3 lemons

1. Put the elderflowers, sugar, vinegar and water into a large jar or bowl. Squeeze the juice from the lemons, cut the rind into thick strips and add to the container. Stir well, cover and leave for 24 hours. Stir or shake occasionally to dissolve the sugar.
2. Strain into sterilized bottles, leaving about 2 inches space, and screw on the caps. Store in a cool, dry, dark place.

Fresh and aromatic borage and mint make a decorative garnish on elderflower champagne.

ELDERFLOWER MILK SHAKE

A "vitality" drink full of protein, vitamins – and flavor.

SERVES 4
2½ cups buttermilk
3 large heads elderflower
2 tbsp. honey
2 eggs, separated
4 tbsp. plain yogurt
ground cinnamon
scented geranium leaves, to decorate

1. Heat the buttermilk with the elderflowers and honey to just below boiling point. Cool and strain – a coffee filter deals efficiently with the minute flowers.
2. Pour the flavored buttermilk over the egg yolks, beating constantly, or mix them in a blender.
3. Beat the egg whites until stiff, then fold them into the buttermilk.
4. Pour the drink into four tumblers, top each one with a spoonful of yogurt and sprinkle it with cinnamon. Decorate with geranium leaves. Serve chilled with a wide "fun" straw.

ROSE GERANIUM BOWL

A nonalcoholic "punch" in which the apple juice is transformed.

MAKES ABOUT 3 PINTS.
3 pints unsweetened apple juice, chilled
⅓ cup granulated sugar
3 lemons
8-12 rose-scented geranium leaves, plus extra to
    decorate
3 tbsp. lemon syrup
ice cubes
1¼ cups carbonated mineral water

1. Put the apple juice and sugar into a pan. Squeeze the juice of two lemons. Cut up the rind and add it to the pan with the geranium leaves. Heat gently until the sugar dissolves, then bring to a boil, stirring occasionally. Boil for five minutes, then set aside to cool.
2. Pour the syrup and lemon juice over ice cubes in a pitcher, strain in the apple juice and pour in the mineral water. Mix well and decorate with thin slices of lemon and geranium leaves.

MARIGOLD COOLER

A nonalcoholic punch, chilled tea spiced with marigold petals, makes a refreshing drink.

MAKES ABOUT 2 PINTS
4 tsp. Earl Grey tea
4 tsp. marigold petals
2 strips thinly pared orange rind
1 pint boiling water
about 8 tbsp. crushed ice
juice of 2 oranges
2 tsp. grenadine syrup
1¼ cups carbonated mineral water, chilled
4 marigold flowers
4 sprigs lemon balm or mint

1. Put the tea, marigold petals and orange rind into a pot, pour in the boiling water, stir well, cover the pot and leave to infuse for one hour.
2. Put the crushed ice into a pitcher, pour in the orange juice and syrup, and strain in the tea. Cover and chill for one hour. Pour in the mineral water and stir. Decorate with the marigolds and herb sprigs.

MINT AND BORAGE CUP

You can make this tingling-fresh herb lemonade with other aromatic leaves. Try varying the recipe with lemon balm or golden marjoram.

MAKES ABOUT 2 PINTS
1 handful fresh mint leaves
4 tbsp. water
juice of 2 lemons
1¾ cups carbonated mineral water, chilled
borage flowers, to decorate
ice cubes, to serve
FOR THE SYRUP
2 tbsp. granulated sugar
1 cup water

1. To make the syrup, put the sugar and water into a pan, heat gently until the sugar dissolves, then bring to a boil, stirring occasionally. Boil for five minutes, cool and chill.
2. Put the mint in a blender or food processor and liquefy with the water and juice of one lemon.
3. Stir together the syrup, mint mixture and juice of the other lemon. Cover and chill for one hour.
4. Stir in the mineral water, decorate with the borage flowers and serve with ice cubes.

INDEX

Numbers in italics refer to illustrations

Picture Credits
Alan Duns: half title, 8/9, 68, 72/3, 90, 91, 92, 97, 132.
Ray Duns: 29, 66, 87, 101, 105, 106, 110/1, 113, 117, 119, 123,
124, 130/1, 135, 137, 146/7, 158.
Laurie Evans: 4/5, 156/7.
Nelson Hargreaves: 13, 14, 16, 16/7, 18, 22, 24 (t, b), 26, 27, 30,
31, 32, 39, 43, 81, 83, 163.
James Jackson: 153.
Di Lewis: 149, 161.
Peter Reilly: 2, 6, 15, 21, 25, 36/7, 38, 40, 44, 45, 49, 50, 53, 54,
57, 58/9, 60, 62, 63, 64, 67, 71, 75, 121, 126/7, 129, 139, 140,
143, 145.

Acknowledgments
The publishers would like to thank Joan Green and Micky
Scammell for making the artificial flowers on pages 105, 119,
123, 124; Clive Ross for making the small roses on page 119;
Dorothy Ross for making the flowers on pages 101, 106, 110/11;
and Sue Trenchard for making the flowers on page 113.

The publishers would also like to thank Ronnie's Fresh Flower
Stall, Berwick Street Market, London, for supplying most of the
flowers photographed in this book, and Albert Weaver, Master
Baker, of Wethersfield, Braintree, Essex, for making the flower
loaf and rolls.